A LOVE FOR FOOD

A LOVE FOR FOOD

RECIPES FROM THE FIELDS AND KITCHENS OF DAYLESFORD FARM

Carole Bamford

daylesford
ORGANIC FARM GLOUCESTERSHIRE

◧ SQUARE PEG

1 3 5 7 9 10 8 6 4 2

Square Peg, an imprint of Vintage,
20 Vauxhall Bridge Road,
London SW1V 2SA

Square Peg is part of the Penguin Random House group of companies whose addresses can be found at global.penguinrandomhouse.com

Text copyright @ Carole Bamford 2020

Photography © Sarah Maingot
Except photographs on pages 6, 11, 28, 44, 60, 90, 135, 146, 150, 151, 202, 209, 352 and 369 by Martin Morrell; 262 and 263 by Britt Willoughby Dyer

Carole Bamford has asserted her right to be identified as the author of this Work in accordance with the Copyright, Designs and Patents Act 1988

First published in the United Kingdom by Fourth Estate in 2013

penguin.co.uk/vintage

A CIP catalogue record for this book is available from the British Library

ISBN 9781529110296

Printed and bound in Italy by L.E.G.O. Spa

Penguin Random House is committed to a sustainable future for our business, our readers and our planet. This book is made from Forest Stewardship Council® certified paper.

Recipe notes:
Oven temperatures are for a conventional oven. If you are using a fan oven, please reduce by 10°C.

Contents

7	*Introduction*
11	On Toast
29	*Sustainability and the Environment*
33	Eggs
45	Soups
61	*The Market Garden*
69	Salads
117	Vegetables
147	*The Creamery and Cheese Room*
151	Savoury Tarts and Pies
183	Fish
203	*The Animals*
209	Meat
245	Puddings
277	*The Bakery*
281	Breads
307	Cakes and Breaks
333	Staples
377	Index
383	Acknowledgements
384	About the Author

The first edition of this book was published in 2013, and while a lot has changed at Daylesford since then, and the farm has grown and evolved; in a way much remains the same. The messages that we were trying to convey with that first book – to celebrate seasonal ingredients and to eat in a way that feels nourishing and balanced for you and for the planet – are as relevant and important today as they were then. And the philosophy and principles on which Daylesford was founded continue to drive and inform everything that we do: to farm and produce seasonal, sustainable, organic food; to nurture the soil and the land that we work here; to live in harmony with nature's cycles; and to try to help and encourage others to live a little more consciously, mindful of their own impact on the health of the earth.

Of course, the world has moved on and changed since then too. The appetite for everything that we have been talking about has also increased and it's been heartening and encouraging to witness the evolution of the conversations around sustainable living and the accompanying shifts in mindset and behaviour. Acts such as shunning plastic bags, opting for a reusable container and trying to address the world's crippling food waste problem are just some of the many things that are now commonplace behaviours – and they are just the tip of the iceberg. Creative and inspiring endeavours are being undertaken by individuals and businesses around the world to try to tackle the climate emergency and help slow the rate at which the earth's temperature is rising. Every day I am encouraged and inspired by the voices of the younger generation. The urgency of the actions taken by young activists bring into sharp relief the state of the crisis we are in and remind us all that we cannot sit by and let this happen. Their collective voices fill me with enormous hope and make me proud to think that what we do here at Daylesford may have played a small part in the swell of that movement.

Anybody who has read my previous book Nurture will be familiar with the story of why and how I started farming organically at Daylesford and will know that I really didn't mean for it to have grown as it has; one thing just led to another. Even its beginning came about after a series of chance encounters. The first happened when we were living and farming at Wootton in Staffordshire. It was the late 1970s; I had not long given birth to my eldest daughter Alice and I was in the garden, pushing her around in her pram. We went to look at the roses I had planted just a few days earlier only to find that they were wilting. Like most other farms at this time we had big fields because all the hedgerows had been pulled out to free up space. I spoke to the farmer next door who explained that farms were being sprayed with Roundup. I didn't even know what Roundup was, but I quickly learned that it was a very powerful herbicide and that the toxins had been carried in the air from the fields and had caused my roses to wilt. As a new mother, I was frightened and horrified, and I knew instantly that we couldn't carry on farming as we were, harming the earth in this way – it wasn't right for the health of my children and it wasn't right for nature and the land.

Not long afterwards I was at an agricultural show where an organic farmer had a small tent. The organic movement was quite a small, niche thing in those days, but I went inside and spoke to the farmer for a few hours. On the way home I remember saying to my husband, 'We can't carry on farming as we are. We are polluting the soil and the environment with those chemicals. For our children and for future generations, we have to stop.' It seemed so simple and obvious

INTRODUCTION

to me: unless you look after the health of the soil, you cannot grow healthy plants and you cannot produce healthy food for humans or for our animals. So we went to see the farm manager, who didn't really believe that switching to organic farming was possible, but I was quite determined.

That was the start. We farmed deer, cattle, sheep and chickens in Staffordshire, then we came down to Gloucestershire in 1992 and again, the first thing we did was to convert the land to become organic. It took a long time for it to clean itself because of the chemicals that had been used previously, but suddenly wonderful things started happening: wild flowers and plants that hadn't been seen before began appearing in the undergrowth. I particularly remember seeing many different kinds of violets emerging. And our soil began changing: it became rich and dark and alive with worms – very different in colour and appearance to any soil that has had chemicals in it.

Organic farming isn't easy – crops fail, the weather interferes with your plans and it's a slower way to cultivate and produce food, but I believe that for the sake of the planet and its future, it is the best way to farm, and my biggest hope is that we will see things change to the point where it becomes possible for everybody to work their land in this way. Current food production systems have brought the world's resources to their knees and we are facing a situation in which over half the world's soil has been stripped of its nutrients and the continued existence of plants, animals and vital pollinators is under threat. If we don't address the wrongs of the past and switch to a form of agriculture that regenerates and nourishes the soil and builds its fertility, we will have very few growing cycles left.

But attitudes around food are changing: people are beginning to understand and appreciate the true value of food and the work that goes into producing it, but also realising the damage that wasting it does to our planet. Food waste is currently the world's third largest polluter, which is hard to fathom and comprehend as it's so unnecessary and preventable. It's something that as a society we vitally need to address – from an individual perspective in our homes as well on a corporate and governmental level.

More and more people do want to know what they're eating and how their food is produced – and they are arming themselves with information, seeking out and supporting the businesses and producers that are behaving responsibly. With organic food, that reassurance and transparency comes from being certified by the Soil Association. The principles that underpin and govern their accreditation criteria mean that food that is sold and labelled as organic must, legally, be traceable from farm to fork. It guarantees that food has been produced with no synthetic fertilisers, herbicides or pesticides – it uses natural fertilisers from plants; requires less energy than industrial farming and holds the welfare of the animals as a top priority. As a farm, it's a source of enormous pride to us that all of our Daylesford products hold that certification. It is a visible, vital indicator to our customers that not only are they purchasing something sustainable and ethical, but also that as a business we are going that bit further and having a positive impact on the planet.

But organic is also a term that has been mistreated and misunderstood, too often used as a marketing tool to convince people to eat something grown out of season, in unnatural conditions, on the basis that it might be better for them. To me that makes no sense either, which is why at Daylesford we prefer to talk about food that is local, seasonal and sustainable.

The food I love to eat is mostly very simple – I like dishes where the ingredients are left to shine rather than those that involve lots of fuss on a plate. And the recipe books I treasured most when I was first married and learning to cook were the kind where you wrote notes in the margin and were full of down-to-earth, nourishing food for the family – they were the starting point and inspiration for this one. Many of the recipes here are old favourites of mine, which have been in my family for a long time and have been passed down by my mother and grandmother, and which I still love today. But plenty have been created by the talented chefs I have been very fortunate to have worked with at Daylesford – many of whom are still with us. John Hardwick has been in charge of the kitchens at Daylesford for over 17 years now, and Gaven Fuller, our head chef, has worked alongside him for over 15. Between them they have created not only many of the recipes in this book, but the food that you enjoy in our cafés, on our deli counters and at our events. You will spot some of those stalwarts and popular dishes among those here. I am not a natural cook and so all of the advice, tips and knowledge that you'll find on these pages is thanks to their wisdom and skill, not mine.

Our intention is that these recipes reflect Daylesford's emphasis on fresh, seasonal, nourishing food – and by that I mean food that feeds body and soul. For me, there is as much nourishment in a hearty, rustic cheese and leek pie as there is in a plate of fresh, flavoursome and seasonal tomato salad or a rich, decadent chocolate brownie. Each has its place in a varied, balanced diet.

Scattered throughout the book are also stories and notes from the farm – insights from the passionate experts I am so lucky to be surrounded by: individually they nurture our livestock, ensure the market garden flourishes, create our artisan cheeses and breads and ensure that we continue to do so in a way that is not only sustainable and kind to the earth, but in a way that gives back to it. I am deeply indebted to them for all that they do to keep Daylesford thriving.

A whole tapestry of people has also helped and supported me right from the beginning, especially Carlo Petrini, who founded the Slow Food Movement. He inspired me so much with his love of food and his support for artisans and sustainable systems. And Patrick Holden, who was then President of the Soil Association, would come down to the farm to advise us on what to do. Again, Daylesford would not be what it is today without their guidance, encouragement and reassurance. Their work is being continued by a new generation of pioneers: children are taking a stand and my hope is that they will continue to reinforce the message that together we can continue to be a force for change.

There is an old saying that I particularly like, which is along the lines of: 'The earth is not a gift from our parents; we are only borrowing it from our children.' I have grandchildren now too, and I often take them around the farm. They love to watch the sheep-shearing and see the lambs. We've put on protective suits to visit the beehives, we walk around the market garden and at Christmas we go to see the turkeys together. The children understand that this is where their food comes from and that makes me very happy. I hope that as you cook your way through this book, you will celebrate your own love for food, but also that its pages may encourage you to reflect on where that food comes from and be inspired to play your part in helping to ensure we nurture the earth that offers up such a delicious and wonderful larder.

ON TOAST

Notes on toast

These recipes are simply a collection of things that are served on charred or toasted bread, and can be made small and daintier for picking up with fingers and serving with drinks or left chunkier, to eat with a knife and fork. They would make a great brunch, sharing plate, starter, lunch or supper.

There are also a couple of terrines, which pair beautifully with chutney, piccalilli or red tomato chilli jam (see pages 343–351 for recipes). Terrines do take a little bit of effort, but they are great things to make ahead when you have friends coming round, as all the work is done in advance. The thing to do is make a big terrine – more than you need in one go – then it will keep in the fridge for the rest of the week, ready to be served on crusty bread or toast for a quick lunch or supper.

Toast is as English as can be, but Italian bruschetta is toast taken to a different level. Bruschetta, at its simplest, is toasted bread rubbed with a little garlic and drizzled with really good olive oil, plus a few flakes of sea salt. Just that. But obviously you can also add whatever toppings you like.

The bread is either grilled (under a medium grill, so you can do it fairly slowly) or preferably griddled, rather than done in a toaster. If it is griddled we also like to put the toast into the oven briefly, so that when you add the oil, tomato or topping, you have brilliant, slightly charred, crispy crusts but keep the softness inside. Not everyone does the oven stage, but for me this is the way to get the best result.

It is crucial to start with the right bread. You want a good, open-textured bread with bubbles through it: sourdough is just right. You can cut it about a centimetre thick and it will be crunchy on the outside, but thick enough to stay good and soft in the centre.

We also make very thin, light, oven-baked toasts, which we use in chicken Caesar salad (page 114) and to serve with a light pâté such as the smoked mackerel one opposite, as it is nice to pair them up with something equally light (and crispy).

SERVES 6

Smoked Mackerel Pâté with Daylesford Toasts

This is a very quick, easy, tasty pâté, traditionally served with light, crunchy toasts. Ours (which we also make for chicken Caesar salad, using seeded bread) are somewhere between thin crostini and classic melba toasts. Our version is made with sourdough bread, very thinly sliced, then brushed with olive oil and crisped in the oven.

700g smoked mackerel fillets, skin removed

200g crème fraîche

2 tablespoons chopped fresh flat-leaf parsley

juice of 1 lemon

a pinch of cayenne pepper

1 small loaf of good bread

3 tablespoons olive oil

8 small handfuls of mixed salad leaves

3 tablespoons French dressing (see page 335)

2 lemons, cut into quarters

sea salt and freshly ground black pepper

Put the mackerel fillets into a food processor and blend until smooth. Scoop out and put into a bowl. Add the crème fraîche, parsley, lemon juice and cayenne and mix together well, then taste and season as necessary. Spoon into six small serving pots or ramekins and put into the fridge to chill for at least 2 hours, or overnight.

When ready to serve, preheat the oven to 180°C/gas 4.

Cut the bread into wafer-thin slices – approximately 2mm thick – and lay them on a large baking tray. Drizzle each slice with a little olive oil and season with a pinch of salt and freshly ground black pepper. Bake for 6–8 minutes in the oven until very crisp and golden brown. Leave to cool.

Toss the salad leaves lightly with the dressing and divide between six plates. Add a pot of chilled pâté and a wedge of lemon, and serve with a basket of the toasts.

SERVES 4

Garden Vegetables with Hot Cheddar Sauce and Salsa Verde Mayonnaise

This is simply about celebrating the bounty of the summer garden on any given day. The vegetables can change as the season goes on and you might want to add baby beetroots when they are young, tender and sweet. We tend to keep the root vegetables just to carrots and radishes, however, so that the overall feeling is light, rather than too earthy. The only thing to consider is that there is no hiding place for the vegetables, so everything you choose has to have real freshness and flavour.

We serve the crudités on a board, with some boiled eggs nestling in among them, and a pot each of hot Cheddar sauce and cold salsa verde mayonnaise.

4 medium eggs

1 bunch of asparagus, woody ends removed

2 bunches of baby radishes, trimmed

1 bunch of baby carrots, trimmed, stalks left on

2 handfuls of sugar snap peas

2 baby gem lettuces, halved lengthways

3 stalks of vine cherry tomatoes (about 20)

5 or 6 spring onions, trimmed

1 handful of pea shoots

4 slices of good bread, preferably sourdough

sea salt

1 quantity of hot Cheddar sauce (see page 341), to serve

½ quantity of salsa verde mayonnaise (see page 336), to serve

Have the eggs at room temperature. Bring a medium pan of water to the boil and add a pinch of salt (this will make the eggs easier to peel). Gently lower in the eggs and simmer for exactly 7 minutes. Take off the heat and rinse immediately in cold water to prevent discolouring, then peel.

Arrange the prepared vegetables on a board or platter. Halve the eggs and add them to the platter, then sprinkle with the pea shoots.

Toast the bread until crisp – either under the grill or on a griddle until both sides are nicely charred. Cut into soldiers and arrange on the platter.

Warm the Cheddar sauce over a low heat, stirring constantly. Once warm (before it starts to simmer), remove it and pour into a small pot. Have the chilled salsa verde mayonnaise in a separate pot and serve both alongside the crudités.

SERVES 4

Broad Bean, Pea, Mint and Feta Toasts

These are really colourful, bright, fresh bruschetta-style toasts that capture early summer and are great to put out on a big board or platter with drinks. By brushing the toast with oil and rubbing it with garlic first, you seal in these flavours but stop the toast being greasy (see page 12 for more on how to get toasts just right).

After you have zested and squeezed the juice from the lemon half, a nice touch is to grill or griddle the other half, at the same time as the bread, until it just chars, then add it to the serving platter.

150g podded broad beans
110g fresh or frozen peas
1 small bunch of fresh mint, finely shredded
juice and zest of ½ a lemon
4 tablespoons extra virgin olive oil
75g feta cheese

4 slices of sourdough bread (about 1cm thick)
1 clove of garlic, cut in half
1 handful of fresh pea shoots
sea salt and freshly ground black pepper

Preheat the oven to 180°C/gas 4. Have ready a bowl of iced water.

Add the beans to a small pan of boiling water. Cook for 30 seconds, then lift out with a slotted spoon and plunge them into the iced water. Add the peas to the boiling water. If they are frozen, cook as before, for just 30 seconds, then drain and transfer to a bowl (if you are using fresh peas, they will need to cook for about 5 minutes to become tender). Lightly crush the peas with the back of a fork.

Once the beans are cold, slip off the outer skins and add the beans to the bowl of peas with the mint, lemon juice and zest, and 2 tablespoons of the olive oil. Crumble in the feta, mix gently and season well.

Either heat a grill or get a griddle pan smoking hot on the hob. Lightly char the bread on both sides, then transfer to a baking tray. Rub each slice of toast with the cut surface of the garlic, drizzle with 1 tablespoon of the oil and season with a little salt and pepper.

Put into the oven for about 6 minutes or until crisp on the outside but still soft on the inside. Remove from the oven, halve each slice of toast and arrange on a board. Top with the broad bean mixture, add a few pea shoots, and drizzle with the remaining olive oil.

SERVES 6

Welsh Rarebit and Chutney

A good Welsh rarebit is all about the balance of strong, mature Cheddar and the spice and heat that comes from mustard and Worcestershire sauce. If you use a subtle-tasting cheese, its flavour will be overpowered by the spice, but if you don't add enough of the condiments, you really just have cheese on toast. I have a particular weak spot for cheese and so like a really strong Cheddar for my rarebit – one that's been aged for a long time.

Some people toast the bread first, then pile the cheese mixture on top and grill it, but I prefer it if it's all cooked in the oven – that way you avoid either soggy or rock-hard toast. You do have to start with well-textured bread as it had to be a good carrier for the cheese mixture; a sourdough is perfect.

The apple and chilli or butternut squash chutneys on pages 346 and 349 go especially well with the rarebit.

The cheese mixture can be kept in the fridge for up to a week if you don't want to use it all in one go.

100ml milk

500g mature Cheddar cheese, grated

2½ tablespoons plain flour

2 tablespoons English mustard

1½ tablespoons Worcestershire sauce

2 eggs, plus 2 egg yolks

a little olive oil, for greasing

6 slices of good bread, preferably sourdough

6 tablespoons chutney (see recipes on pages 346–349)

6 small handfuls of mixed salad leaves

3 tablespoons French dressing (see page 335)

sea salt and freshly ground black pepper

Preheat the oven to 190°C/gas 5.

Put the milk and grated cheese into a pan over a gentle heat and let the cheese melt slowly. Gradually add the flour, mustard and Worcestershire sauce and stir constantly until all are incorporated and you have a smooth sauce. Remove the pan from the heat and allow to cool, then beat in the whole eggs and yolks and season.

Grease a large baking sheet with a little olive oil.

When ready to serve, spread a nice thick layer of the cheese mixture on each slice of bread and arrange on the prepared baking sheet.

Put into the oven for about 10 minutes, until golden brown on top and slightly crispy underneath.

Serve with a good dollop of chutney and the salad leaves, tossed in the dressing.

MAKES 12 SLICES

Ham Hock Terrine with Piccalilli

Ham hocks – the 'ankle' of the pig – are often forgotten about, but they are so flavoursome. Our chefs like to serve this terrine in the autumn, with the nicely matured piccalilli that was made from the summer glut of vegetables in the market garden and some toasted sourdough.

You need to start this a couple of days before you want to serve it to soak the ham hocks and rest the finished terrine.

3 ham hocks	juice and zest of 1 lemon
2 heads of garlic	2 tablespoons capers, drained
3 sticks of celery, roughly chopped	2 tablespoons Dijon mustard
1 leek, roughly chopped	6 gelatine leaves
2 carrots, roughly chopped	12 tablespoons piccalilli (for homemade see page 343)
1 big bunch of fresh flat-leaf parsley, stalks and leaves chopped separately	sea salt and freshly ground black pepper

Soak the ham hocks in water for 24 hours, then drain and rinse them. Put them into a large pan and add enough fresh cold water to cover the ham by 5–7cm. Add the vegetables and parsley stalks. Bring to a simmer and cook for about 3 hours, until the ham is very tender, topping up with boiling water as necessary so that the liquid level doesn't drop below the top of the ham.

Lift out the hocks (reserving the cooking liquid) and flake the flesh, removing all the fat and gristle. Put into a bowl with the parsley leaves, lemon juice and zest, capers and mustard. Mix well and season.

Put the gelatine leaves into a bowl of ice-cold water until soft (the water must be cold or the gelatine will dissolve). Take out, squeeze, and put to one side.

Line a terrine tin with clingfilm and spoon in the ham mixture. Strain the ham cooking liquid and measure 1.2 litres of it into a pan (top up with water if necessary). Bring to the boil, then take off the heat and stir in the gelatine leaves, continuing to stir until they are completely dissolved. Leave to stand for 15 minutes, until tepid, then pour over the terrine so that all the ham is just covered. Put into the fridge to chill overnight. When ready to serve, slice the terrine – this will be easiest with a sharp, serrated knife – and serve with the piccalilli.

Venison Terrine with Tomato and Chilli Jam

There is something about the pairing of juniper and venison that evokes winter woods and log fires. If you make the tomato and chilli jam (see page 351) when the tomatoes are at their best in the summer, the flavours will have softened out and mingled nicely by autumn, ready to serve alongside chunky slices of the terrine.

Venison is one of the leanest meats, with negligible fat content, but in order to keep the terrine moist you do need some fat, hence the pork belly.

125g butter, plus a little extra for greasing the terrine
2 shallots, finely chopped
2 cloves of garlic, chopped
200ml port
700g venison shoulder, chopped
550g pork belly, chopped
1 teaspoon freshly grated nutmeg
1 teaspoon ground ginger
3 juniper berries, crushed
70ml double cream
3 eggs, beaten
70g fresh breadcrumbs
450g sliced smoked streaky bacon
sea salt and freshly ground black pepper
12 tablespoons red tomato chilli jam (see page 351), to serve
12 slices of good bread, preferably sourdough, toasted, to serve

Preheat the oven to 180°C/gas 4.

Melt the butter in a medium pan, add the shallots and garlic and cook gently over a low heat for 5 minutes, until slightly softened but not coloured. Add the port, bring to a simmer and bubble until reduced by half. Remove the pan from the heat and leave to cool.

In a food processor, using the pulse button, or with a hand mincer, mince the venison and pork together and transfer to a large bowl. Add the cooled onion mixture along with the spices, cream, eggs and breadcrumbs and season well. Before you put it into the terrine, you need to check the seasoning. Do this by frying a tiny piece of the mixture in a hot frying pan until the meat is cooked. Taste it, then season again accordingly.

Grease a terrine with a little butter and line it, widthways, with the bacon slices. Make sure each slice butts up against the previous one (you don't need to overlap the slices, or the bacon layer will be too thick). The slices need to overhang the edge of the terrine by about 7–8cm on either side, so if they are not long enough, use 2 slices, butted head on in the centre of the terrine.

Carefully spoon in the terrine mixture and fold the overhanging slices of bacon over the top. This time, you do need to overlap the ends slightly, as the top of the terrine will be subjected to the highest heat and the slices will shrink and pull apart otherwise. Cover the top of the terrine with foil, put it into a roasting tin and fill the tin with enough hot water to come halfway up the outside of the terrine, creating a bain-marie.

Put into the oven for 1¾ hours, until cooked through. To test, run the cold tap over a metal skewer, so that it is very cold. Insert into the centre of the terrine, leave it there for 5 seconds, then quickly remove it and carefully press it to the back of your hand. If the skewer is still cold, the terrine is not ready. If the skewer is warm, it is ready. If the skewer is about to burn your hand, then the terrine is overcooked. Leave to rest and cool down for 1 hour so that the heat from the outside of the terrine transfers to the centre, resulting in even cooking throughout. Once cold, put into the fridge for 24 hours.

To serve, slice the terrine and serve with the red tomato chilli jam and a basket of the toasted bread.

Hot Dorset Crab on Toast

SERVES 4

This makes a lovely, simple, but indulgent light meal, with a really satisfying contrast in textures: a little crustiness from the grilled cheese on top, then you crack through to the gooey crab and finally hit the crispness of the toast.

Always buy fresh, unpasteurised, hand-picked crabmeat. By combining some of the more flavoursome brown meat with the very delicate white meat, you have a mixture that can hold its own inside the creamy, slightly mustardy mayonnaise and under the cheese topping.

300g fresh white crabmeat

200g fresh brown crabmeat

80g mayonnaise

2 teaspoons English mustard

juice of ¼ of a lemon, plus 1 whole lemon, quartered, to serve

a little oil, for greasing

4 slices of good bread, preferably sourdough

2 tablespoons finely grated Parmesan cheese

2 tablespoons finely grated Cheddar cheese

1 tablespoon chopped fresh flat-leaf parsley

2 tablespoons extra virgin olive oil

4 small handfuls of mixed salad leaves

4 tablespoons French dressing (see page 335)

sea salt and freshly ground black pepper

lemon wedges, to serve

Preheat the oven to 180°C/gas 4. Lightly oil a baking tray.

In a bowl, mix together the crabmeat, mayonnaise, mustard and lemon juice, and season well.

Preheat the grill or heat a griddle pan, and lightly toast the bread on one side only. Arrange on the prepared baking tray, toasted side down, then generously spread each slice with the crab mixture and sprinkle with a little of each cheese. Put into the oven for about 8 minutes, until the cheese melts and turns golden brown. Take out of the oven, sprinkle with the chopped parsley, drizzle with extra virgin oil and add a twist of black pepper. Serve with the salad leaves, tossed with the dressing, and a wedge of lemon.

SERVES 6

Sardines on Toast with Tomato, Basil and Caper Relish

This is one of my favourite suppers when I want something comforting and simple. The sardines can be served without the relish for a meal that can be put together in moments but I love the kick that it offers. If you want more heat in the relish, use a hotter chilli.

6 slices of good bread, preferably sourdough

12 fresh sardines, butterflied

3 tablespoons olive oil

1 clove of garlic, cut in half

extra virgin olive oil for drizzling

3 large vine tomatoes

6 small handfuls of rocket leaves

sea salt and ground black pepper

For the tomato caper relish:

3 tablespoons olive oil

1 large red onion, thinly sliced

a pinch of coriander seeds

a pinch of fennel seeds

50ml white wine vinegar

40g caster sugar

1 teaspoon finely chopped long red chilli

5 large red vine tomatoes, chopped

4 tablespoons capers

juice and zest of 1 lemon

1 handful of fresh basil, leaves roughly chopped or torn

Heat the olive oil for the relish in a pan, then put in the red onion and both seeds and cook gently over a low heat until the onions have softened. Add the vinegar, sugar, chilli, tomatoes and 200ml of water. Bring to a simmer and continue to cook gently, stirring occasionally, until the tomatoes have broken down and all the liquid has evaporated. Stir in the capers, lemon juice and zest, and the basil, remove and leave to cool, then taste and season as necessary.

Preheat the oven to 180°C/gas 4. Heat a grill or get a griddle pan smoking hot on the hob. Lightly char the bread on both sides, then transfer to a baking tray and put into the oven for about 6 minutes.

Meanwhile, rub the sardines with olive oil and season. Put under a hot grill or, preferably, cook on a hot griddle for 2 minutes on each side, until lightly charred and cooked – check by gently pressing the flesh with the back of a fork and if it starts to flake it is ready.

Rub the toasted bread with the cut side of the garlic and drizzle with olive oil. Slice the tomatoes and divide between six plates. Put a slice of toast on top with some tomato relish, followed by 2 sardines, some rocket leaves, drizzle with extra virgin olive oil and some black pepper.

Sustainability and the Environment
Tim Field

After I finished my degree, I was working as an environmental consultant when the chance came up to join Daylesford. They had already been farming organically here for years, but Carole Bamford wanted to go beyond that, to strive to become completely self-sufficient and also to share our experiences of doing so with others.

We've always been ambitious with our targets, working across the business to ensure that every department and area was doing everything it could to function in as sustainable way as possible, but in the past few years we've become particularly focused on specific goals: becoming zero waste, generating our own clean energy and championing a food system fit for the future.

Striving for zero waste

Composting plays a huge role in managing the food waste side – we compost all of the fruit and vegetable trimmings from the kitchens and our ready-meal production unit. It is a wonderful cyclical process, in which nothing travels further than a few hundred yards: the produce goes from the market garden to the kitchens and then the scraps go to the compost and back on to the gardens again. But we also try to prevent any waste in the first place by ensuring that we balance the different areas of the business. We have a nose-to-tail policy with our livestock carcasses, so the bones will be boiled and made into broths, then used in anaerobic digestion to produce energy; the less popular cuts of meat will be used to make the stews, tagines and casseroles in our fresh ready-meal range, while the popular prime cuts go to our butchers' counters. Similarly, any gluts of vegetables from the market garden that can't be sold in the farm shops will be sent to the fermentary where they will be preserved as nutritious, fermented foods. And any surplus food that we can't sell in our shops by its 'best before' date will be donated to the Felix Project, a charitable organisation with whom we've partnered that distributes food to some of London's most vulnerable and in need.

Our packaging strategy has been aimed at removing any single-use items from the business, increasing the amount of recycled material content, finding innovative materials that can replace plastics and also focusing on giving materials a second life – ensuring that they are recyclable, compostable or reusable so that they are kept within a cycle and we don't generate anything that would leave this loop and need to go into landfill.

Most recently, we've also launched our zero waste pantry, which ties all of these principles together. The dedicated space within our farmshop allows customers to bring their own reusable vessels to stock up on storecupboard essentials – everything from pasta and rice to olive oil, vinegar and laundry detergent. It allows them to buy what they need to avoid any waste and, of course, creates no new packaging. It's really heartening to watch people shopping in this way, as it shows that the appetite for shopping consciously is there and people are willing to change their habits to come together and play their part in addressing the environmental crisis we are facing.

Clean energy

We now have over 2,000 solar panels at the

farm site which, in high summer, can generate all the electricity we need to power the farmshop, bread ovens, the creamery, dairy and the refrigeration, all from the sun, and we export any excess back on to the National Grid. In 2016 we installed a biomass boiler, a green source of energy that burns locally sourced wood to produce all our heat requirements; we also have an energy capsule that catches the heat above the log-fired oven in the kitchens and from behind the fridges, and heats the water for the kitchens; and our own sustainable rainwater system, which will eventually be used for all of the farm, so that one day we can be completely off-grid in terms of energy and water.

Farming sustainably

Carole always reminds me that two things are necessary for sustainable food production: healthy soil and the bees. Healthy soil is the essence of everything. If you look after it and encourage the microbial and earthworm activity, you end up with a rich, light, aerated soil, which can retain moisture but won't get compacted and heavy, and that doesn't need chemicals.

Bees, too, have an incredibly important role to play in a sustainable system. We have two beekeepers: one based in the market garden and one producing estate honey, in small quantities, but in which you can really taste individual flowers or herbs at certain times of the year. In the market garden, the beekeeper has a policy of minimal intervention, so we let the queens and their colonies leave their hives and go where they need to: perhaps a hollow in a tree or a chimney pot or, with any luck, a vacant hive where we encourage them to take up residence.

Organic systems also rely on beneficial predator insects that control pests. We could plant all sorts of wildflower meadow strips to attract them, but actually we don't need to, since the diverse way that Jez organises his planting does more than anything else to attract the 'good' insects naturally because throughout the year there are hundreds of different fruits and vegetables all flowering or producing seedheads or leaf material at different times.

This holistic view of the ecosystem is at the core of the way we farm but a recent initiative takes this a step further. Agroforestry is a means of managing combinations of trees, crops and livestock so that not only do you get better results from your crops, but you enhance the welfare of your animals and give back to nature in other ways too. Our project is focused around improving the welfare of our hens.

We have planted 800 fruit trees that run in rows along the lengths of half our hen fields. Alongside them are rows of native British alder trees that not only act as a windbreak for the fruit, they also offer shelter for the hens and provide woodchip to be used as fuel in our biomass boiler. Alder trees also draw nitrogen from the air into the soil to increase its fertility naturally. In the other half of the fields, we've planted kale. The kale will be grown the year after the hens have roamed the fields. The mobile hen houses will be moved across to the next half of the field, leaving behind a rich, fertile soil, perfect for nourishing the growing kale. The following year the vegetable will be left to go to flower and seed. The seed feeds the birds and the flowers provide essential pollen and nectar for bees and other pollinators, a source they can access throughout the year, which is particularly vital during the more barren periods of the year when food is scarce. It's a system that not only increases the productivity of those 30 acres of land, it has benefits to wildlife, sequesters carbon from the atmosphere, locking it into the soil where it is needed, and also contributes to mitigating the effects of climate change. The planting of in-field trees helps to slow the overland flow of torrential rainfall, helping water to infiltrate into soils rather than race away to the rivers and cause flooding downstream.

The wetlands

I remember on the day I arrived at Daylesford, Richard Smith took me around, and there was a soggy wet patch in one of the fields by the river with lots of weeds, and he said he wanted to do something with it. Wetlands used to be an important part of our environment, but many have been lost through the intensification of farming, so it's been a source of pride to see ours thrive again. Wetlands act as a natural buffer against extreme weather conditions: they absorb water in times of flood and release it in times of drought, allowing us to manage and survive conditions such as these which, as we've seen recently, are sadly an increasing occurrence. They also provide an important habitat for wildlife.

What excites me most, though, about the sustainability efforts here at Daylesford, is the work we do in education and in sharing knowledge, in particular the food projects we do with children through the Daylesford Foundation. We work with schools, helping teachers to set up organic gardens, bringing children on to the farm and engaging them in where their food comes from. The Foundation also set up an online platform for farmers – Agricology – a forum in which farmers come together to share their own experiences and sustainable practices in an effort to support and encourage each other. The site now has an audience of over 150,000 farming professionals, which I find really encouraging.

Our biggest hope with all that we do with our initiatives and efforts here is really about inspiring others. If we can arm our customers, other farmers and growers, as well as the next generation, with the knowledge and tools they need to grow, shop and work sustainably, we can hope to challenge the status quo and continue to spread awareness of our belief that farming and food production systems that are founded on agroecological principles can and will feed the world.

Notes on eggs

Everyone has different ideas on how to achieve the best boiled egg, but the way the chefs cook them at Daylesford is to start with the eggs at room temperature, so that they have less chance of cracking. They lower them into boiling water – not cold – as this allows you to time them more accurately, from the moment they go in.

For a medium-sized soft-boiled egg that you can dip soldiers into, boil for 5 minutes; for a medium egg that has a vibrant, slightly soft yolk – boil for 7 minutes. If you want a really hard egg, for example, for egg mayonnaise, boil for 10 minutes. If you are peeling the eggs, then as soon as you take the pan off the hob, put it under the cold tap until the eggs are cool, which will prevent a grey-blue line forming around the yolks. A pinch of salt in the cooking water helps stop the whites clinging to the shells and makes peeling easier.

At Daylesford we have various breeds of hen, including the Blue Legbar, whose eggs are instantly recognisable because they have shells that are naturally coloured from blue through to turquoise to a pale khaki green. We probably get fewer eggs per bird than many farmers, but our hens have over-sized ranges to run around and forage in, and spacious houses to shelter in at night. Then every few months the houses and runs are moved to new, fresh pasture, leaving behind naturally manured ground for our market garden crops. If a hen is outside the majority of the time, foraging and scratching around, eating grass and clover, plucking at insects, as it would do if you kept it in your back garden, then that hen will have had a happy, healthy quality of life and the egg will have a rich golden yolk and a real flavour.

This situation is sadly quite different to many systems – even free-range. If you have 15,000 birds in a henhouse, you can encourage them to go outside all you like, but they are not going to be bothered to climb over 300 of their friends to get to the pop hole and go free ranging; they have food and water inside, why do they need to go anywhere else? As a result of the hen's inactivity and diet, the average supermarket egg has a pale yolk and tastes quite bland and neutral.

Bubble and Squeak with Fried Egg

SERVES 4 AS BRUNCH

At home my family always makes some iteration of bubble and squeak on Boxing Day, but it's a good recipe for any time you have had a roast dinner and have potatoes and vegetables left over.

Whether you are starting from scratch or using up cooked vegetables, the key is to chop the vegetables and colour them in a pan before mixing them with mashed potato to get the roasted flavour and crunchy edges that make bubble and squeak so appealing (even if you have roast potatoes to add to the mix, you need a little mash to bind everything together).

You can make the bubble and squeak by just piling the mixture into the pan and moving it around so that it is hot all the way through and you have lots of crispy brown surfaces. However, if you shape the mix into cakes, you can make them in advance and chill them in the fridge for a while, which also helps to set and hold the cakes together when you fry them.

3 large potatoes, such as Sante or the red Romano, peeled and cut into 6 pieces

½ a medium swede, peeled and cut into 6 pieces

about 6 tablespoons rapeseed oil

60g butter

1 medium onion, sliced

1 clove of garlic, chopped

70g Savoy cabbage, thinly sliced

150g Brussels sprouts, chopped

1 head of broccoli, separated into florets

2 tablespoons chopped fresh flat-leaf parsley

250g fine polenta

4 eggs

sea salt and freshly ground black pepper

good tomato ketchup, to serve (see page 361)

salad leaves (optional), with a little French dressing (see page 335), to serve

Bubble and Squeak continued

Put the potatoes and swede into a pan and cover with cold, slightly salted water. Bring to the boil, then turn down the heat and simmer until just tender. Drain in a colander, then put into a large bowl and mash.

Heat 2 tablespoons of the oil and half the butter in a non-stick frying pan and put in the onion, garlic, cabbage, sprouts and broccoli. Cook gently for about 15 minutes, until golden brown and softened. Add to the potato and swede, together with the parsley, and season well.

Form the mixture into 8 cakes, transfer to a plate or tray and let them cool, then put them into the fridge for 1–2 hours.

When ready to cook, preheat the oven to 130°C/gas 1.

Have the polenta ready in a shallow bowl. Dip each cake in it to coat, shaking off any excess.

Heat 2 more tablespoons of the oil with the rest of the butter in a non-stick frying pan and fry the cakes in two batches, until golden brown on both sides and hot in the middle, putting the first batch into the oven to keep warm while you cook the second. Add a little more oil if necessary.

Meanwhile, heat the rest of the oil and fry the eggs to your liking.

Serve one egg on top of two bubble-and-squeak cakes per person, with a good dollop of tomato ketchup and, if you like, some salad leaves, tossed in French dressing.

Pan Haggerty with Mustard, Egg and Caper Mayonnaise

Potatoes, onions and cheese are a great combination, and pan haggerty is the traditional Northumberland version of it, to which we add bacon, then serve it with some salad leaves tossed in French dressing and a mustard and caper mayonnaise. The tartness of the dressing and the capers counters the richness of the cheese and bacon.

700g potatoes (we use Maris Piper or Desiree), peeled

50g butter

1 medium onion, finely chopped

75g smoked streaky bacon (or pancetta), chopped

2 cloves of garlic, finely chopped

150g Cheddar cheese, coarsely grated

2 tablespoons chopped fresh flat-leaf parsley

olive oil, for shallow frying

4 small handfuls of mixed salad leaves

a splash of French dressing (see page 335)

For the mayonnaise:

2 eggs

150g mayonnaise

50g wholegrain mustard

50g capers

2 tablespoons chopped fresh flat-leaf parsley

sea salt and freshly ground black pepper

Preheat the oven to 180°C/gas 4.

Leaving the peeled potatoes whole, put them into a pan and cover with slightly salted cold water. Bring to the boil, then turn down the heat to a simmer and cook for about 15 minutes, until the potatoes are still firm and slightly undercooked. Drain in a colander and when cool enough to handle, grate them, coarsely, into a large mixing bowl.

While the potatoes are cooking, bring a medium pan of water to the boil and add a pinch of salt (this will make the eggs easier to peel). Gently lower in the eggs and simmer for exactly 10 minutes. Take off the heat and rinse immediately in cold water to prevent discolouring, then peel the eggs and grate them roughly into a bowl. Add the mayonnaise, mustard, capers, parsley and seasoning and mix well. Keep in the fridge while you cook the pan haggerty.

Pan Haggerty continued

Melt the butter in a small pan, then add the onion, bacon and garlic and cook gently until the onion is soft but not coloured. Take off the heat and cool a little, then mix into the grated potato, together with the cheese and parsley, and season with salt and plenty of pepper.

Form the mixture into 4 cakes. Heat the oil in a non-stick frying pan that will transfer to the oven, then put in the cakes and gently brown them on each side. Transfer to the oven for about 10 minutes, until crisp on the outside and soft and cooked in the middle.

Serve each pan haggerty with a handful of salad leaves, tossed in the dressing, on top, and a spoonful of the mustard caper mayonnaise on the side.

SERVES 6

Corned Beef Hash Cakes

This is a twist on the classic New York brunch dish, and though it can be made with tinned corned beef, that tends to be a lot fattier. So the proportions here are really designed for the homemade corned brisket on page 236. You can eat the brisket hot one day, then shred what is left for this hash; it's great for a lazy, late Sunday morning.

The Daylesford chefs make their own brown sauce and if you want to do the same there is a recipe on page 402.

500g potatoes, peeled and quartered

750g corned beef (see page 236), brought to room temperature then chopped

100g butter

2 onions, finely chopped

¼ fresh red chilli, deseeded and finely chopped

a dash of Tabasco sauce, to taste

1 tablespoon Worcestershire sauce

1 tablespoon chopped fresh flat-leaf parsley

2–3 tablespoons plain flour

2 tablespoons olive oil, plus a little extra for frying the eggs

6 eggs

sea salt and freshly ground black pepper

6 tablespoons brown sauce, to serve (for homemade, see page 362)

Preheat the oven to 180°C/gas 4.

Put the potatoes into a pan and cover with cold, slightly salted water. Bring to the boil, then turn down the heat and simmer until just tender. Strain then put into a large bowl and mash. Mix the corned beef with the mashed potato. Heat half the butter in a small pan, then put in the onions and cook gently for 5 minutes, until softened. Mix into the potato and beef, along with the chilli, Tabasco, Worcestershire sauce, parsley and some salt and pepper.

Have the flour ready in a shallow bowl. Form the beef mixture into 6 cakes and dust each with a little flour, shaking off the excess.

Heat the remaining butter with the olive oil in a frying pan and cook the cakes, in batches, over a medium heat until hot and golden brown on both sides. Place on a baking tray in the oven for 5 minutes, to heat right through to the middle. Meanwhile heat some more olive oil in a separate frying pan and fry the eggs. Serve each cake with a fried egg on top and brown sauce on the side.

SERVES 4

Rita's Baked Eggs and Onions

Over the years we have had a number of cooks from Barbados in the Daylesford kitchens, and this is a dish made famous by the wonderful Rita. No matter who makes it, it will never taste quite like Rita's.

It is pure comfort food, especially if you serve it with wholemeal or basmati rice (about 500g, cooked and drained).

8 eggs	1 tablespoon Dijon mustard
175g butter	50g Gruyère cheese, grated
4 large onions, thinly sliced	120ml cream
80g plain flour	20g Parmesan cheese, grated
850ml milk	sea salt and freshly ground black pepper

Preheat the oven to 180°C/gas 4.

Bring a medium pan of water to the boil and add a pinch of salt (this will make the eggs easier to peel). Gently drop in the eggs and simmer for exactly 10 minutes. Take off the heat, rinse the eggs immediately in cold water to prevent discolouring, then peel.

Melt 75g of the butter in a separate pan. Add the onions and cook over a low heat with a lid on for about 10 minutes, until softened, then season well, remove from the heat and leave to cool.

Melt the remaining butter in a small pan over a low heat. Add the flour, whisking to a smooth paste, then cook gently for 5 minutes. Gradually whisk in the milk and stir until thickened and smooth. Add the mustard and Gruyère, and when the cheese has melted add the cream. Taste, season as necessary, then take off the heat.

Divide half the onions between four ovenproof shallow dishes or cast-iron skillets (you can also use one large shallow ovenproof dish). Slice 4 of the hard-boiled eggs and arrange on top, then spoon over a third of the cheese sauce. Repeat the layers, finishing with the remaining sauce, and sprinkle with the Parmesan. Bake in the oven for about 15 minutes (if using a single dish it may take longer), until hot and golden brown on top. Remove from the oven and leave to stand for 5 minutes to cool slightly and allow the flavours to merge before eating.

Notes on soups

Making soup is one of our favourite ways to celebrate the market garden and inspiration for our soups often begins with Jez, who will come into the kitchen laden with whatever is in abundance there.

Unless you are making something like a minestrone, which is all about using chunks of a medley of different vegetables, the secret is to keep to two or three key flavours, so that rather than ending up with something indistinct, you really taste every ingredient.

The smaller you can chop your vegetables the better – about 1cm – rather than leaving them in big chunks, as this allows the soup to cook more quickly and keep its flavour, goodness and colour. Also, if the pieces are to stay whole, rather than be blended, as in a minestrone, when you put your spoon into the bowl you want to scoop up a good selection in one go. Although it might seem tedious, try to cut the pieces into similar sizes, too, so they cook evenly.

Soup needs a good stock. If you just use water you don't get the same depth of flavour. Vegetable stock, especially, is such an easy thing to make at home and a great way to use up that couple of carrots, lone leek or wedge of butternut squash. You can make plenty, then freeze it in bags or ice-cube trays (see pages 372–374 for stock recipes).

Soups made from earthy root vegetables like beetroot and carrot quite often need a little sharpness to give them a lift. More often than not a dash of lemon juice is a good way of bringing out your flavours (in any soup, really) even more: you won't taste it specifically, but it will add a little extra zing. And when you season your soup with salt and pepper, do this gradually, especially with the salt. Taste a spoonful, add a little cautious seasoning, stir the soup, leave it for a minute or so, taste it and season lightly again if necessary. Do this a few times until you are happy.

SERVES 6

Celeriac and Apple

Celeriac and apples are harvested at the same time in autumn and they make a classic combination. You want a tart, full-flavoured apple, such as a Cox, or look for Bountiful, which Jez calls a nearly-cooking variety, which sweetens by October to suit dishes like this.

300g good, tart eating apples, see introduction above

juice of 2 lemons

50g butter

500g celeriac, peeled and chopped

1 litre good chicken or vegetable stock (see pages 375 and 372)

150ml double cream

sea salt and freshly ground black pepper

Peel, core and chop the apples and put them into a bowl with the lemon juice.

Heat the butter in a large pan (one that has a lid) over a low heat until melted, then add the celeriac, put the lid on the pan and cook over a medium heat for 10 minutes, without letting the celeriac colour, stirring occasionally.

Add the apples and lemon juice, put the lid back on and cook for a further 5 minutes.

Add the stock and the cream and bring to the boil, then turn down to a simmer for about 10 minutes, or until the celeriac and apples are soft.

Blend, taste and season as necessary before serving.

SERVES 4

Chilled Pea and Mint

This is the simplest of soups to celebrate summer. It is lovely to make when you have fresh peas; however, all the podding does take time and there is absolutely nothing wrong with a pea that has been frozen as soon as it is picked – in fact, when it comes to soups and purées, frozen peas can be preferable because they give you that really intense bright, summery green colour.

600ml good vegetable stock (see page 372)
750g frozen or fresh peas
200ml double cream
juice of ½ a lemon
2 tablespoons chopped fresh mint leaves, plus extra to garnish
sea salt and freshly ground black pepper
a little extra virgin olive oil, to serve (optional)

Bring the stock to the boil in a medium-sized pan and add the peas. Bring back to the boil, then remove from the heat and stir in the cream, lemon juice and mint.

Blend until smooth, then taste and season with salt and pepper.

Cool and put into the fridge until chilled. Serve in chilled bowls and, if you like, drizzle with a little olive oil and scatter with the mint leaves.

Chilled Tomato, Cucumber and Fennel

SERVES 6

Everything in this summer garden soup is raw, but as well as the bright, fresh flavours, the pleasure comes from the different textures. Some of the tomatoes are blended, while the cucumber, fennel, pepper and celery are chopped finely and mixed through to lend a nice crunch. You want really sweet cherry tomatoes. Gardener's Delight are easily accessible, but throughout the summer Jez sends the kitchens heritage tomatoes in every shape and colour from the market garden.

800ml tomato passata

150g cherry tomatoes roughly chopped, plus 4 ripe red vine cherry tomatoes, chopped small (about 0.5cm)

6 tablespoons olive oil

2 tablespoons balsamic vinegar

juice of ½ a lemon

¼ of a fresh red chilli, deseeded and finely chopped

1 clove of garlic, crushed

1 teaspoon sugar

1 cucumber, deseeded and chopped small (about 0.5cm)

1 fennel bulb, chopped small (about 0.5cm)

½ a red pepper, deseeded and chopped small (about 0.5cm)

½ a yellow pepper, deseeded and chopped small (about 0.5cm)

½ a green pepper, deseeded and chopped small (about 0.5cm)

1 stick of celery, chopped small (about 0.5cm)

2 tablespoons chopped fresh flat-leaf parsley

3 tablespoons shredded fresh coriander

a little extra virgin olive oil

sea salt and freshly ground black pepper

In a large jug, combine the passata, 150g cherry tomatoes, olive oil, balsamic vinegar, lemon juice, chilli, garlic and sugar with 500ml water. Blend until smooth.

Mix the rest of the chopped vegetables and the 4 red vine cherry tomatoes together in a large bowl, pour the blended tomato mixture over them and put into the fridge to chill for 1–2 hours.

Just before serving, chill six bowls. Stir the parsley and coriander into the soup, taste and season as necessary, then serve, drizzled with a little olive oil.

SERVES 6

Beetroot, Bacon and Crème Fraîche

This is one for the autumn, when the big winter beetroots are around. Their earthiness combines beautifully with the smokiness of the bacon and the crème fraîche, which has a little more acidity than cream, so adds a touch of sharpness and gives the soup a lift, drawing out the flavour of the beetroot rather than adding richness.

2 tablespoons vegetable oil

150g smoked streaky bacon, finely chopped

2 onions, finely chopped

500g raw beetroot, peeled and chopped

juice of ½ a lemon

1 sprig of fresh thyme, leaves only

1 litre good chicken stock (see page 373)

100g crème fraîche

sea salt and freshly ground black pepper

Heat the oil in a large pan (one that has a lid), add the bacon and sauté until golden brown.

Add the onions, reduce the heat, put the lid on the pan and cook for about 10 minutes until softened.

Add the beetroot and continue to cook, with the lid on, over a medium to low heat for another 10 minutes, stirring to prevent sticking.

Add the lemon juice, thyme leaves and stock and bring to the boil, then reduce the heat and simmer with the lid on for about 25 minutes.

Add the crème fraîche and bring back to the boil, then remove from the heat. Blend, taste and season as necessary before serving.

SERVES 8

Chicken (or Turkey), Ginger and Vegetable Broth

This is a broth that feels as though it is doing you good. It is a handy recipe for making a day or so after you have roasted a chicken, as you can turn the carcass into stock and shred any slivers of leftover meat into the soup. It is also perfect to make with turkey on Boxing Day. Our chefs put it on the menu in January and people love it after the indulgence of Christmas.

You can put in the vegetables in any proportion you prefer – and if you don't have them all it doesn't matter, but the greater the mixture the more flavoursome and healthier the broth will be.

If you want to make turkey stock, crush the carcass with your hands, to break the bones, and put them into a large pan with 2 large white onions, peeled and halved; 2 carrots, peeled and halved; 2 heads of unpeeled garlic, just cut in half; a small bunch of fresh thyme; and a tablespoon each of peppercorns and sea salt. Cover with cold water, bring to the boil, then reduce the heat to a simmer and skim off any grease and scum on the surface. Then simmer gently for 2–3 hours.

- 2.5 litres good chicken stock (see page 373) or turkey stock
- 400g mix of shredded red onion, carrots, cabbage, leeks, celeriac or celery and kale
- about 2cm fresh root ginger, finely chopped
- 500g cooked chicken or turkey meat, torn into thin strips
- 3 tablespoons chopped fresh flat-leaf parsley
- 2 tablespoons chopped fresh chervil
- sea salt and freshly ground black pepper

Bring the stock to the boil, add all the prepared vegetables and the ginger, and simmer for about 5 minutes, until all the vegetables are soft. Add the chicken or turkey and bring back to the boil, then take off the heat.

Taste, adjust the seasoning as necessary, and sprinkle with the herbs before serving.

SERVES 6

Butternut Squash, Honey and Sage

This soup is quite rich – so you don't need to serve enormous bowls of it. It's smooth and silky, with a great colour and has just the right element of sweetness.

Of the readily available squashes, you can't beat butternut, which is full of flavour and has the right texture and vibrant colour, whereas some varieties have disappointingly dull-looking flesh. And if you've got heritage squashes to choose from, Crown Prince is also good, but avoid pumpkin as the flesh tends to be more watery.

80g butter	150ml double cream
1kg butternut squash, peeled, deseeded and chopped	juice of ½ a lemon
	75g honey
1 teaspoon sea salt	2 tablespoons finely chopped fresh sage
1.1 litres good chicken stock (see page 373)	freshly ground black pepper

Melt the butter in a large pan (one that has a lid) over a low heat, then add the squash and salt. Cover with the lid and sweat for 10 minutes over a low heat, until softened but not browned.

Add the stock and the cream, turn up the heat and bring to the boil, then turn down to a simmer for 5 minutes, or until the squash is cooked through and soft.

Add the lemon juice, honey and sage, bring back to the boil, then remove from the heat. Blend, taste and season as necessary before serving.

SERVES 6

Leek and Potato

A very British, very comforting classic that will always be a part of the Daylesford repertoire – especially as leeks are a vegetable that thrives in our market garden. Around March/April they are young and tender and perfect for this soup, combined with floury potatoes, such as King Edward or Cara, which will break down well in the stock.

It is a very simple soup, but because of that it can either be really satisfying, or bland and gloopy if you use too much potato over leek or leave it cooking for too long, so that it stews and looks rather grey and you miss the slightly peppery sharpness of the leeks coming through. Some people insist on only using the white of the leeks, but we don't believe in throwing away the green parts, which also help to keep the soup looking fresh and green. So we use plenty of both colours, chopped finely, and if the potatoes are also cut small, the leeks will hold their colour in the short time that the potato takes to cook.

Our chefs often add some grated mature Cheddar to it before serving – a couple of tablespoons per bowl.

80g butter

2 onions, peeled and chopped

3 cloves of garlic, sliced

2 teaspoons fresh thyme leaves

200g potatoes, peeled and chopped (see above)

500g leeks, finely chopped

1.4 litres good vegetable stock (see page 372)

150ml double cream

sea salt and plenty of freshly ground black pepper

mature Cheddar cheese, grated, to serve (optional)

Melt the butter in a large pan (one that has a lid) over a low heat, then add the onions, garlic and thyme leaves. Put the lid on the pan and cook gently for about 5 minutes, until the onions have softened, stirring occasionally.

Add the potatoes and cook for a further 5 minutes with the lid on, then add the leeks and continue cooking slowly, still with the lid on, until the vegetables have softened. Add the stock and the cream, bring to the boil, then turn down to a simmer and cook for 10 minutes.

Blend, taste and season as necessary before serving, with grated Cheddar if you like, and an extra twist or two of black pepper.

SERVES 6 AS A LIGHT MEAL

Ten Vegetable and Two Grain Minestrone

You can make a version of minestrone all year round, with different combinations of vegetables – though it is good to keep to the base of carrot, onion, celery and courgette. The recipe below is one for late summer/autumn, when the swede and squash are in season, but in high summer you can replace these with fresh peas, broad beans and green beans – it doesn't have to be exactly ten vegetables.

- 3 tablespoons pearled spelt
- 3 tablespoons pearl barley
- 1 tablespoon olive oil
- ½ a medium white onion, chopped
- 1 clove of garlic, finely chopped
- raw meat from 2 small chicken thighs, chopped
- 1 small cooking chorizo (about 100g), chopped
- ½ a small carrot, chopped
- ½ a stick of celery, chopped
- ¼ of a small swede, chopped
- about 70g butternut squash or pumpkin, chopped
- 1 tablespoon tomato paste
- ¼ of a fresh red chilli, finely chopped
- ½ teaspoon smoked paprika
- 1 tablespoon plain flour
- 1 teaspoon fresh thyme leaves
- 1.3 litres good chicken stock (see page 417)
- 500g tinned chopped tomatoes
- ¼ of a small leek, chopped
- ½ a small courgette, chopped
- 2 kale or cabbage leaves, thinly shredded
- a dash of red wine vinegar
- ½ teaspoon sugar
- sea salt and freshly ground black pepper

Put the spelt and pearl barley into a pan and cover with cold water. Bring to a simmer and cook for about 30 minutes, until the grains are tender, then drain in a colander and keep to one side.

Heat the olive oil in a large pan (one that has a lid), put in the onion and garlic, season with a little salt and cook for about 5 minutes over a low heat, until the onion is softened but not coloured. Add the chicken, chorizo, carrot, celery, swede and squash, put the lid on and continue to cook for a further 5 minutes.

Add the tomato paste, chilli, paprika, flour and thyme leaves and cook for 2 more minutes, then add the chicken stock and tomatoes. Bring to a simmer, then add the rest of the ingredients and the spelt and pearl barley and cook for 5 more minutes, until all the vegetables are tender. Taste and adjust the seasoning, adding plenty of black pepper.

The Market Garden
Jez Taylor

Growing fruit and vegetables organically can be something of a lottery, and you seem to forever be saying, 'That was a funny old season'. Each year is different – you're at the seasons' mercy and there's no way of predicting what might happen. Crops could be early or late, magnificent or spoilt by some freak of weather. There is always some new challenge, so you are constantly discovering and adapting.

In the twelve years since I arrived at Daylesford we have extended the market garden from eight to thirty acres, because ultimately we are trying to be as self-sufficient and diverse in our growing as it is possible to be. We want to concentrate on interesting, often heritage crops that have to be grown and harvested by hand and that suit the soil and landscape here. But that is what growing, and especially organic growing, is all about: understanding a particular piece of ground, learning over time what grows best there; and nurturing the soil, preserving the microorganisms, so that you leave the land in a good state to endure seasonal extremes.

In reality what we have at Daylesford isn't naturally the kind of prime, horticultural growing land that the big specialists in East Anglia have. There it is all flat fields, very consistent terrain and quite peaty, rich soils, so it is easy to plan, plant, manage, hoe and harvest mechanically with all their specialist kit. We have land that isn't flat, that has quite a lot of wind blowing through; we have mixed soils and we harvest by hand.

The farm is on the site of an ancient river bed and so there are river pebbles everywhere. It was a big, meandering river, so there are patches of field that are very sandy, others quite sticky clay, and there are significant dips which can flood and become boggy. Then on the slopes around, we have the traditional limestone shale normally associated with the Cotswolds. If you have waterlogged ground, a lot of the goodness will be washed out, much of the microbial, earthworm activity is knocked out and, if you keep going at it, rotovating it or driving on it, it becomes compacted and suffers even more, which of course is the antithesis of the whole ethos of looking after the soil.

So when you have ground that can be quite vulnerable, you have to find your niche and celebrate and emphasise the crops that thrive. Of course we can grow things like carrots, parsnips and potatoes well here, but we don't want to do things just because we can; we want to go to town on the crops that we are really good at, like specialist leaves and herbs, heritage tomatoes, soft fruit, such as gooseberries, blackberries, jostaberries, plums and particularly strawberries; leafy things like kale, spinach and chard, or purple sprouting broccoli; and the whole allium family: onions, garlic and leeks. We are brilliant at leeks. In a good year, we have 40–50,000 of them that we will harvest over a good nine months.

When I arrived here there was an asparagus bed, but it is hard to control the slugs, asparagus beetle and perennial weeds such as couch grass and creeping buttercup – and it only really makes sense if you are a specialist. We like to celebrate local food as well as what we can grow ourselves – so it is better for us to sell beautiful English asparagus from the nearby Vale of Evesham in the farmshop when it comes into season, than battle to grow some of our own.

What we are doing here is exactly what I

THE MARKET GARDEN

would do myself if I had my own smallholding – just on a much bigger scale. I was born on Jersey, where my father worked for the Ministry of Agriculture, helping to develop outdoor tomatoes and early potatoes. Then my parents bought an old farmshop in Cleeve Prior, near the Vale of Evesham, and tried growing everything from sweetcorn to raspberries, before they stumbled across the idea of making hanging baskets and turned the farm into a highly successful hanging basket nursery. I was always interested in growing and food but, because I was good at science, for a mad moment I thought I wanted to be a food technologist, until I started studying the subject at Reading University and it seemed to be mainly about incorporating cheap ingredients into processed food and advertising it well. So I went back to horticulture.

After I finished my degree, I rented two walled gardens near Reading for ten years where I grew specialist salad leaves and vegetables for restaurants and farmers' markets. I developed a box scheme and set up my own business making cider and apple juice in Cleeve Prior, then selling it in Reading as well as doing a bit of tree surgery and willow work. When I was growing up, my father once made a load of apple juice, which was kept in the garage and went fizzier and then sour, because we couldn't drink it all. It was a formative moment for me because it taught me you could do something amazing with unwanted apples.

There used to be around 2,000 acres of apple, perry pear and plum orchards in Victorian times in the Vale of Evesham, and though so many have been grubbed up and lost, some of the small farms still have fruit trees. So I would go around them all and say, 'Can I have your apples?' and they would either give them to me for free or trade them for juice or cider. Eventually I became quite snobbish and only wanted to get my fruit from old, wild orchards. I also started grafting fruit trees, and so my legacy in Cleeve Prior is some 300 cider apple trees planted around my parents' nursery.

When the opportunity came up to be head gardener at Daylesford in 2008 I was doing a consultancy for the Eden Project in Cornwall, helping to develop a market garden there, and the timing was just right. It doesn't matter whether you are growing on a small scale or a big one, the opportunity to grow organic, diverse and seasonal food, hands-on, using age-old skills; to build your own team around you and be part of a traditional mixed farming system, with animals, the creamery and the bakery is one that every grower would seize. And it is great to work closely with chefs whose skill is in pulling off brilliance with simple fresh produce and who appreciate the value of turning gluts into preserves to use for the rest of the year.

And then, a year after I started at Daylesford, by some sort of serendipity, an orchard just happened to become available a couple of miles up the road, planted with different cider apple varieties and with its own cider-making shed and press. I sold up my cider business in Cleeve Prior, and now make local Cotswold cider for Daylesford. In a day at the orchard we can make 1,000 litres. We leave it for at least ten months and once it has fermented to cider, we bottle it and add a small amount of sugar – the equivalent of about a teaspoon per bottle – which balances the acidity. And it is carbonated, which is the way most people want to drink cider.

Making cider is such a simple thing to do,

and is a great way of maintaining a strong local tradition and of using apples that would otherwise go to waste to create a drink that's characteristic of the area. I think cider goes with everything and use it instead of white wine in my cooking.

Lessons from the seasons

Managing an organic market garden is all about the diversity and safety of your food supply. Inevitably you are vulnerable to fluctuations in the weather, the temperature, too much or too little rain, but when you grow as many different crops and varieties as you can sustain, if one fails, you always have something else to fall back on. That way you don't suffer 'funny old seasons' in the way that specialists can do.

Diversity also means that terrible weather conditions for one crop can create a window of opportunity to work on something else. One year, for example, we had the best tomatoes ever. We grow around 3,000 plants in 40 different varieties, many of them heritage: black, green, yellow, orange, stripy; bell-shaped, pear-shaped, plum-shaped, and in that summer we harvested about 4 tons partly as a result of an awful, rainy strawberry season, when we couldn't pick the berries in July because it was too wet. Instead we used the time to take off all the leafy side-shoots from the tomato plants in the tunnels so, without these sapping energy from the plants, all their efforts went into flowering and the subsequent, fantastic fruit. Bad seasons can teach you good lessons.

We regularly take on Soil Association apprentices and trainees. It is great to be able to show young people a rich, diverse and fascinating food culture, and teach skills that are in danger of being lost. We never have idle hands. In agriculture it is very easy to become a cog in the machine of a big operation producing for the supermarkets and doing something mind-numbingly repetitive every day, but here there is always something different happening and new things to learn.

Before Christmas a few years ago the whole farm was covered in a foot of snow, so there was very little that we could do in the gardens. But we have a little willow plantation, so we fashioned big wreaths from willow and foraged around the estate for leaves and berries. The wreaths were so successful that now we make them every year. Once the cider is made, we have the wreath-making . . . and so the cycle goes on.

Growing crops is all about food for me. Of course it is, you might think, but if you were working in a big agricultural packhouse day in, day out, you might not make that same connection. All the things I grow I want to cook and eat – especially the whole squash and pumpkin family. We concentrate on the best culinary ones that are valued for their tasty flesh, in soups, roasted in chunks on their own or in a risotto, rather than the more spectacular Halloween varieties. The secret with squash is to store them in nice warm, airy conditions, then you can keep them right through to April/May. In an ideal world I would like a nice specialist squash and pumpkin store, similar to the cheese room, where I could keep the humidity out.

A love for leaves

What I love to grow and eat most, though – and what I am really known for – are unusual salad leaves. In my family, salad is a meal in itself: the base will be mixed

leaves, chopped up a little, then lots of bean sprouts, which are very easy to produce. I have a kilner jar next to the sink and I put some mung beans, alfalfa or green lentils in the bottom, fill the jar with water, then leave them to soak overnight. The next day I close the jar, but without the rubber seal, so there is a gap for the water to run out. Then I lay the jar on its side, so the water can drain into the sink. I just flush the jar two or three times a day with fresh water and let it drain away, because the only thing that is not good for sprouts is to let them sit in a container full of water. Within a few days I'll have a jarful of sprouts.

Every year we increase the number of salad leaves that we grow: outside in summer, and in the polytunnels in winter – though even into January we can still be harvesting 40 per cent of salad leaves, such as mustard, rocket and endive, outside; as well as those in the tunnels, all of which are unheated, but give natural warmth and protection.

When you grow organically you always have to establish a four-year rotation, so that you don't have the same crop in the same location more than once every four years. This is what helps improve the structure of the soil, lowers the risk of disease and aids pest control. Like all the crops that we grow, the leaves are organised in families, then we can create a rotation of brassica leaves, mustards and rocket; lettuce; endive and herbs.

The key to growing winter salad leaves is to get the plants in the ground in the tunnels early so they have time to develop a good root system. The growing season is principally the period of time between the clocks changing, from the end of March to the end of October. If you suddenly think, 'winter salads, must do something,' in mid-September, it is too late, but if you plant in mid-August, you can keep harvesting the leaves for six months, cutting a few leaves from each plant, and leaving behind the little rosettes in the centre, from where new leaves will grow. Then in April, the leaves come out in all but one of the tunnels (which we keep for delicate salads like rocket and mizuna). Within three weeks we will have moved from winter production to summer production and done all the tunnel planting of summer herbs like parsley, basil and coriander, and fruiting crops, such as tomatoes, chillies and cucumbers. I like small, more European-style cucumbers, which we have learned to grow really well, and they produce little yellow star-shaped flowers, smaller than courgette flowers, which we harvest for the chefs to use as garnishes.

Because we can never cultivate right to the edge of the tunnels, we use up what would otherwise be wasted space by building raised beds, where we grow micro-leaves like Red Frills mustard and really peppery land cress, with leaves that look a bit like crayfish tails and which the chefs love. Around late February we have the greatest variety of leaves of all, everything from butterhead lettuce to radicchio and Japanese mustards like mizuna and mibuna.

I think that a salad needs a touch of bitterness to get the tastebuds going, so I love endive, but not everyone is so keen on bitter flavours. That intensity is a response to UV rays, so the great thing about winter endive is that it doesn't develop the bitter taste as much and is actually pretty sweet.

A sense of place

Every year we invest in more fruit bushes and trees, the most recent being rows of cherry trees, as I wanted a fruit for the kitchens to use that would be ripe after the strawberries have finished and before the blackberries and raspberries. Raspberries are a tricky fruit because picking them is time-consuming and tedious as the berries are very soft and bunched together; strawberries are a lot easier.

People love our strawberries, because we are one of a handful of growers that raise the plants outside in the ground. Ninety per cent of strawberry farms use growbags on tabletops in tunnels, whereas all our plants – around 22,000 of them – are grown in the soil and we have six different varieties, which means that the season is extended naturally and the picking is spread out from early June to late July. That's the theory, but the reality is that if you get a cold, wet, season, of course that doesn't work and everything in the garden comes through at the same time. In a diverse market garden like ours there is always something else that can be done on the days when you can't pick, whereas if you are a specialist grower, you suffer much more.

For me, the perfect strawberry shouldn't be all sweetness – you need some acidity, but mainly it has to be ripe. It sounds obvious, but if you pick a strawberry when it is orange, it will last for a week in a display cabinet, but will never become truly ripe. When we pick ours, they are properly red, and will only last for about two days.

Our varieties are Honeoye, Alice, Fenella, Symphony, Christine and Cambridge Favourite, which I used to think was a great strawberry, very disease-resistant, but it can be quite soft – it's good for jam, though. Our favourite these days is Honeoye, which comes through early and is just the right size. It gets redder and sweeter, producing fruit which stays compact, even in a wet season, for the whole of June, before the likes of Symphony and Fenella take over in July. Beyond the Summer Solstice Garden – where we give demonstrations in growing and cooking, and some of our other skills, such as willow work – and beyond the composting area, are the propagation tunnels where we painstakingly record all the details of every seed that we sow.

Whenever possible, when you are trying to garden sustainably, you save seeds and we have had success with some varieties of peas and climbing and borlotti beans. Plants like the fancy frilly mustards are also good to let go to flower and save the seeds, but seed-saving isn't always straightforward. Squash and courgettes, for example, are very promiscuous and can cross-pollinate with other plants, which might result in 'rogue' types. So we buy many of our heritage varieties from a small company who specialise in old varieties of seeds that grow particularly well in our climate.

The world of propagation is all about nurturing plants slowly from seed and gradually moving them through various stages of being kept warm and covered up until they arrive in the environment in which they end up growing so that it isn't a shock to them. The propagation tunnels are exciting places. It doesn't matter how many years you have been growing fruit and vegetables, the process is still really rewarding. I love how on a cold February day you sow the seeds of spring onions, chillies, peppers, tomatoes and herbs, and over the space of a weekend the tiny plants will suddenly appear.

SALADS

Notes on salads and leaves

Substantial salads that are either starters or meals in themselves are a huge feature of the kitchen menu at Daylesford. They are built up around seasonal produce from the market garden, and while some of them take a little bit of work, others are very simple.

Salads love herbs: lots and lots of them, so even though the recipes might suggest half a bunch, don't hold back – a big flurry of freshly chopped herbs, some gently folded in and the rest scattered over a salad, is what gives that wonderful burst of freshness, flavour and colour from the first forkful.

Jez also has a few tips on growing and using salad leaves: 'The great thing about winter salads, in particular, is that they are relatively easy to grow at home and you can keep them going throughout the cold months from around November all the way through to March/April, under a cold frame or a small polytunnel. Just keep harvesting them cleanly with a sharp knife, from the outside of the plant, so that you leave behind the smaller leaves in the plant rosette, from which more will grow. I don't recommend kitchen windowsills though, as the environment can be a bit extreme, with sunshine or heat from the cooker, and the plants are likely to get hot and stressed as well as leggy where they reach out towards the light.

I don't think there is such a thing as too many different leaves in a salad, but the key is to get a good mix of colour, texture and shape – and to put in something peppery and spicy, something bitter and something with some juiciness and crunchiness, bulked out with more neutral leaves. Remember, once you dress your leaves, the flavour will calm down, but too much of any strong leaf and a salad can become more challenging than enjoyable. I put together different combinations of leaf in summer and winter.

For winter: I would start with salad brassicas, such as mizuna, mispoona, tatsoi, Chinese cabbage hearts, young tender spinach, kale and bull's blood beetroot. Corn salad (or mâche) and winter purslane (claytonia) give a bit of juiciness.

Any of the mustards: red giant, green-in-the-snow, red frills and golden frills mustard as well as salad and wild rocket, and land cress, will give a hit of peppery spice. Some of these, like the land cress, can be really spicy and dominant, though, so you need some winter lettuce, such as oak leaf and broad leaf, to neutralise the spice.

For bitterness I would put in endive or a radicchio – the red and white striped Treviso type.

For summer: I still have my pick of some of the salad brassicas for pepperiness, but there will be no tatsoi, Chinese cabbage, kale or green-in-the-snow mustard. I will include sorrel when it's at its leafy best in April and May. All the other mustards and rockets are still around too, and I like to use a lot of chard – when we have spinach and chard out in the fields in May and June, I will thin out the young tender leaves – nothing bigger than 6cm – for salads as they add colour and juiciness.

The key difference, though, between a winter and a summer leaf salad is that in summer there is a much wider choice of lettuces. And you do want more of these in summer: so look for little gems, cos and variegated leaves like batavia. Around April/June I like to add pea shoots. We take the shoots from the mangetout growing in polytunnels until about May and then we move on to those from the peas out in the fields, so there is a natural progression throughout the season.'

Broad Bean, Bulgar Wheat and Herb

SERVES 6

The first broad beans are harvested around the second week of June and will be small and very tender. Although the vivid green colour looks fantastic if you slip off the skins once you have cooked them, this can be quite fiddly – so if you don't have time, don't worry, the flavour will be just as good. By the end of the season the beans will be quite big and easier to skin, but their flavour will be beginning to be slightly bitter. So the optimum time is mid-season, when the beans are medium size, quite tender, not bitter, and not too hard to skin.

225g bulgar wheat

400g freshly podded baby broad beans

½ a cucumber, cut lengthways, seeds scraped out, flesh cut into 1cm cubes

½ a medium red onion, finely chopped

½ a plump red chilli, finely chopped

3 cloves of garlic, crushed

1 small bunch of mint, leaves roughly chopped

1 small bunch of flat-leaf parsley, roughly chopped

juice and zest of 2 small lemons

4 tablespoons extra virgin olive oil

sea salt and freshly ground black pepper

Half-fill a large saucepan with water, stir in 1 teaspoon of salt and bring to the boil. Add the bulgar wheat, return to the boil and cook for 8–10 minutes, until tender. Drain well in a sieve under running water until cold, then tip into a large serving bowl.

Have ready a bowl of iced water. Half-fill a medium pan with water and bring to the boil. Add the broad beans and return to the boil. Cook for 2 minutes, then drain and put them into the iced water. Leave for 5 minutes, then slip off their skins and add the beans to the bulgar wheat.

Add the cucumber to the bowl of wheat and beans, along with the onion, chilli, garlic and herbs. Add the lemon juice and zest, olive oil and plenty of black pepper, and then mix until all the ingredients are thoroughly combined. Taste, adjust the seasoning as necessary, and serve.

Grilled Peaches, Spelt, Peas, Rocket and Mozzarella

This is the sort of dish I love to serve at a relaxed summer lunch. It evokes long summer days, the kind when you can set a table outside, and put out a big bowl for family or friends, perhaps alongside some plates of prosciutto and other cured meats. This salad has evolved from the grilled peaches that we used to serve whenever luxurious, creamy burrata cheese was available from Italy, into a celebration of summer garden produce, including green beans and peppery rocket, combined with buffalo mozzarella. You need ripe peaches, but not so soft and juicy that they won't stand up to being grilled.

The spelt in the salad isn't meant to be substantial; it is little more than a handful of grains, just to add a scattering of texture and nuttiness.

4 tablespoons pearled spelt

4 ripe but firm peaches

2 teaspoons olive oil

150g green beans, trimmed

140g fresh or frozen peas

1 small bunch of fresh basil

6 small handfuls of rocket leaves

zest of 1 lemon

5 tablespoons French dressing (see page 335)

500g buffalo mozzarella

sea salt and freshly ground black pepper

2 tablespoons extra virgin olive oil, to finish

Half fill a medium pan with water, add ½ teaspoon of sea salt and bring to the boil. Stir in the spelt and return to a simmer. Cook for 20–30 minutes, or until the spelt is just tender, stirring occasionally, then drain through a sieve under running water until cold.

While the spelt is cooking, cut the peaches in half and remove the stones. Cut each peach half into three and rub with the olive oil. Preheat a griddle pan over a high heat (or alternatively heat the grill).

Grill the peach wedges for 2–3 minutes on each side, until lightly charred, then remove with tongs and transfer to a plate to cool.

Have ready a bowl of iced water. Half fill a medium pan with water and bring to the boil. Add the beans, bring back to the boil and cook for 1 minute, then lift out with a slotted spoon (leaving the water in the pan) and put into the bowl of iced water. Leave for 5 minutes, then drain.

Return the water in the pan to the boil and add the peas. Bring back to the boil again, cook for about 5 minutes if fresh, or take off the heat after 30 seconds if frozen, then immediately drain in a colander and rinse under plenty of running water until completely cold. Tip into a large wide serving bowl or platter and add the drained beans and the spelt.

Scatter with the basil, rocket and lemon zest. Add the peach wedges, pour over the dressing and toss gently together. Season to taste and toss gently. Tear the mozzarella into bite-size pieces, scatter over the salad and drizzle with the extra virgin olive oil.

Asparagus, Spelt, Peas and Mint

SERVES 6

English asparagus will be around for about 6 weeks, and at the beginning of the season it will be tender enough to use every part of the spears; but as the weeks go on it will become a little more woody, so you will need to take off the lower, white parts. As in the previous recipe, the spelt here just adds a little texture, but you can leave it out if you like.

300g pearled spelt

2 bunches of slender asparagus

300g fresh or frozen peas

1 small bunch of fresh mint, roughly chopped

1 small bunch of fresh flat-leaf parsley, roughly chopped

5 tablespoons extra virgin olive oil

juice of 2 large lemons (roughly 5 tablespoons)

1 handful of pea shoots

sea salt and freshly ground black pepper

Half fill a medium saucepan with water, add ½ teaspoon of sea salt, and bring to the boil. Stir in the spelt and return to a simmer. Cook for about 20 minutes, or until the spelt is just tender, stirring occasionally, then drain through a sieve under running water until cold. Drain well, then tip into a large serving bowl or platter.

While the spelt is cooking, cut the asparagus into 4cm lengths on the diagonal. Half fill a large saucepan with water and bring to the boil. Add the asparagus and bring back to the boil. Immediately take off the heat (you want the asparagus to be quite crisp), drain in a colander under running water until the asparagus is completely cold and add to the spelt.

Cook the peas in the same way if frozen (if fresh they will need to simmer for about 5 minutes, until tender) and, when cold, add them to the spelt and asparagus, with the mint, parsley, olive oil and lemon juice, and toss well together. Season to taste and toss very lightly again. Scatter the pea shoots over the salad, fold in gently and serve.

Notes on tomatoes

Jez has a top ten list of tomatoes – some easy to find, some more unusual, heritage ones – which are all so popular that we also sell the plants so people can grow them in their own gardens. For a tomato salad I would choose either all cherry tomatoes of different colours, just halved, so you can keep their integrity; or all big, beef tomatoes, which I would slice – these meatier tomatoes are best combined with chunky ingredients, so they are fantastic in a classic tomato, mozzarella and basil salad.

Purple Russian – purple, plum-shaped, sweet and fleshy, with low acidity and a soft skin. These come through quite early and the plants are really vigorous.

Black Cherry – not quite black, but a really dark purple, they are similar in flavour to Gardener's Delight (below) and look fantastic.

Green Zebra – this dark green and yellow stripy fruit is often picked early for its colour, but at the beginning of the season its flavour hasn't developed properly and it can be quite sharp. You need to resist picking it until it has gone the distance and ripened properly (though it will stay green) when it will become sweet and meaty.

Tigerella – really sweet, rich and tangy – and great to look at: orangey-red, with a yellow stripe.

Sungold – these are a hybrid cordon variety and one of the sweetest little orange tomatoes you can find.

Gardener's Delight – everyone knows this one: probably the most popular red cherry tomato, but full of flavour.

Marmande – this one is a slow burner, which often doesn't ripen until August, but is worth the wait. It is a big, ribbed, bulbous tomato, with really meaty flesh, great for slicing and for sauces.

Costoluto Florentina – a heavily ribbed bulbous red tomato; a classic heritage type that looks very architectural on a plate.

Blue Fire starts off bluey green and matures to a smoky dark red with speckles; they're reminiscent of something from space! They have a great meaty flavour when ripe.

Orange Banana are actually a large plum shape; they are sweet and juicy and the flesh resembles mango flesh when chopped.

Tomatoes and Feta with Mint and Lemon Dressing

SERVES 4 AS A LIGHT LUNCH

This is another classic Daylesford salad: a very simple, light, Mediterranean-style dish for the height of summer that relies on beautiful produce – salty cheese and sweet, big, juicy beefsteak tomatoes. I love tomatoes that have warmth and sunshine in them. If you can find heritage Marmande tomatoes – which look like little red ribbed pumpkins, those are some of the best to my mind.

Leave the shredding of the mint leaves until the last minute, to prevent them from turning black.

300g mayonnaise

juice of 2 lemons

8 large tomatoes, preferably heritage, plus a handful of cherry tomatoes, halved

400g feta cheese, drained

1 small bunch of fresh mint

plenty of extra virgin olive oil

sea salt and freshly ground black pepper

Mix the mayonnaise, lemon juice and a couple of twists of ground black pepper in a small bowl until thoroughly combined – it should have a pourable consistency.

Slice the large tomatoes thinly and arrange on a serving platter. Season with a little more freshly ground black pepper and salt.

Crumble the feta into small chunks and scatter over the cherry tomatoes.

Drizzle the salad with the lemon mayonnaise. Finely shred the mint leaves and scatter over the top, finish with a good slug of olive oil, and serve.

SERVES 6

Tomato and Sourdough with Red Pepper, Onion and Basil

This is based on the famous Italian salad from Tuscany, panzanella – one of those dishes that was originally concocted to use up bread that was a few days old. However, instead of adding stale bread, the chefs lightly toast cubes of sourdough in the oven. This is a salad for the height of summer and is only as good as the tomatoes you use. Colourful, heritage varieties, full of flavour, to look out for include Green Zebra, Tigerella and Purple Russian.

2 thick slices of sourdough bread, cubed (about 2cm)

about 8 good flavoursome tomatoes

1 cucumber

1 medium red onion, finely sliced

1 large red pepper, deseeded and chopped (about 1.5cm)

3 cloves of garlic, crushed

1 small bunch of fresh basil, leaves torn

1 small bunch of fresh flat-leaf parsley, roughly chopped

4 tablespoons extra virgin olive oil

2 tablespoons red wine vinegar

sea salt and freshly ground black pepper

Preheat the oven to 190°C/gas 5.

Scatter the cubes of bread over a large baking tray in a single layer. Bake in the oven for 7–8 minutes, until dry and only lightly coloured, then leave to cool.

Cut each tomato into 8 and put into a large bowl. Cut the cucumber lengthways in half. Scrape out the seeds with a teaspoon and chop (about 1.5cm). Add to the tomatoes, along with the red onion and red pepper. Scatter the toasted bread on top and add the garlic, basil and parsley. Season and toss lightly.

Mix 3 tablespoons of the oil with the vinegar, pour over the salad and toss well. Tip gently into a serving dish and drizzle with the remaining oil. Leave at room temperature for around 20 minutes to let all the flavours develop and soak into the bread, then serve.

SERVES 6

Cold Rose Veal with Tuna and Caper Mayonnaise

This is based on the Italian classic *vitello tonnato*, and is one of my favourite things to eat. It is one of the first recipes I learned when I got married. It is a dish you can make all year round, but I think it lends itself well to a long, lingering summer lunch outdoors, when you can serve it alongside a selection of cold dishes such as the tomato and feta with mint and lemon dressing (page 81), the crunchy 'chopped' vegetables (page 89) and the chicken Caesar salad (page 114).

750ml white wine

½ a medium onion, sliced

½ a medium carrot, chopped

3 cloves of garlic

1 teaspoon fresh thyme leaves

900g rump of veal

250g mayonnaise

100g tinned tuna, drained

juice of ½ a lemon

1 tablespoon chopped fresh flat-leaf parsley

2 tablespoons small capers, chopped

4 small handfuls of rocket leaves (or micro-leaves, if you can find them)

sea salt and freshly ground black pepper

extra virgin olive oil, for drizzling

Put the wine, onion, carrot, garlic and thyme into a large pan with 1.5 litres of water and bring to the boil. Put in the whole rump of veal and simmer gently for 15 minutes. Take the pan from the heat and leave the veal to cool down in the cooking liquid (stock).

Meanwhile, in a bowl, mix together the mayonnaise, tuna, lemon juice, parsley and capers, and season well.

Lift out the cooled veal from the stock and slice it very thinly. Divide between six plates and spread a little of the mayonnaise over each slice.

Scatter the leaves over the top and finish with a drizzle of extra virgin olive oil and some salt and pepper.

SERVES 6

Raw Slaw with Chilli, Soy and Ginger Dressing

This recipe has been a staple at the salad counter in our farmshop for several years. It has evolved over that time, but remains the recipe that people most request. It was first introduced by Kuttiya, our Thai chef, and is based on the classic noodle dish, pad Thai, but without the noodles – the idea is that the strips of vegetables should resemble these. The only criteria is that all the raw vegetables must be crunchy, so things like kohlrabi, mixed beansprouts, fresh, firm cucumber and radishes can all be added. The salad doesn't need salt and pepper, as the soy sauce provides saltiness and the chilli gives peppery heat. This is one of the few dishes on the kitchen's menu that is quite showy, as the chefs use a mandoline to shred the vegetables but you can use a very sharp knife to cut them into strips instead.

3 medium carrots

3 raw medium beetroots

¼ of a small red cabbage

¼ of a small white cabbage

1 small red pepper, deseeded and finely sliced

½ a medium red onion, finely sliced

1 small bunch of fresh mint, leaves roughly chopped

1 small bunch of fresh coriander, leaves only

25g cashew nuts, lightly toasted in a dry frying pan and roughly chopped

For the soy, chilli and ginger dressing:

100ml dark soy sauce

100g clear honey

juice of 4 large limes

1 tablespoon caster sugar

2 cloves of garlic, crushed and finely chopped

2cm piece of fresh root ginger, peeled and finely chopped

1 hot red chilli pepper, deseeded and finely chopped

75g cashew nuts

To make the dressing, put all the ingredients, apart from the cashew nuts, into a small pan. Bring to a gentle simmer, stirring constantly, and cook for 2 minutes, then remove from the heat and leave to cool.

Put the cashew nuts for the dressing into a food processor, then add the cooled dressing and blitz until smooth. Keep to one side.

If you have a Japanese mandoline, use the attachment that cuts into curls to shred the carrots, otherwise use a flat mandoline with a medium attachment to shred them lengthways into spaghetti-like strips or cut them into very fine, long julienne strips with a sharp

Raw Slaw continued

knife. Put the shredded carrots into a large bowl. Peel and shred the beetroots in the same way and add to the carrots.

Remove any damaged outer leaves from the cabbages as well as the central core, and discard. Shred the cabbages very finely and add to the bowl, along with the pepper, red onion, mint, coriander leaves and toasted cashews, and toss lightly.

To serve, drizzle with the dressing, toss lightly and tumble on to a serving platter.

SERVES 6–8

Griddled Courgettes and Pine Nuts in Yoghurt and Mint Dressing

This is quite a chunky salad for the end of summer especially, when courgettes are at their best. It makes a good light starter as well as an accompaniment to grilled meat. Serve it at room temperature.

5 medium courgettes

4 tablespoons extra virgin olive oil

4 plump cloves of garlic, sliced into very thin slivers

juice of ½ a lemon

1 small bunch of fresh mint, leaves roughly chopped

150ml natural yoghurt

50g pine nuts, lightly toasted in a dry pan

1 small bunch of fresh basil, leaves roughly torn

1 small red onion, very finely sliced into rings

sea salt and freshly ground black pepper

Preheat the oven to 200°C/Gas 6.

Get a grill or griddle pan hot. Cut the courgettes in half lengthways and lay them cut side up under the grill or cut side down on the griddle pan. Cook for 4–5 minutes, until lightly charred.

When the courgettes are cool enough to handle, cut them into chunks on the diagonal, roughly 2cm thick, then put them into a large bowl and toss with 2 tablespoons of the olive oil. Scatter over a large baking tray in a single layer and roast in the oven for about 8 minutes, until tender.

Tip the hot courgettes carefully into a serving dish and add the garlic and lemon juice. Toss well together and season. Leave to cool a little.

To make the dressing, put the chopped mint into a bowl and add the remaining olive oil and the yoghurt. Taste and season as necessary, mix well, then spoon over the warm courgettes. Scatter with the pine nuts, basil and red onion and serve straightaway.

SERVES 6 AS A LIGHT STARTER

Crunchy 'Chopped' Vegetables

This recipe does involve a lot of chopping, as everything has to be diced as small as you can manage – little more than the size of a matchstick head. There isn't much pleasure in chomping your way through great chunks of raw carrot, radish, cauliflower, but a bowlful of light, crunchy little cubes of mixed vegetables, united by the dressing that can be forked up easily, makes a lovely refreshing starter.

2 large carrots

1 large cucumber, cut in half and deseeded

4 sticks of celery

6 red radishes

2 sweet red peppers, cut in half and deseeded

1 medium head of broccoli

1 medium cauliflower

1 large red onion, finely chopped

250ml French dressing (see page 335)

juice of ½ a large lemon

sea salt and freshly ground black pepper

To cut the carrots into tiny dice, first slice them lengthways very thinly, using a mandoline or a sharp knife, then slice these strips again, into long thin batons, around 3mm wide. Finally cut the batons into dice of around 3mm. Put into a large bowl. Repeat with the cucumber and add to the bowl. Cut the celery, radishes and peppers into similar-sized dice and, again, add to the bowl.

Thinly slice the broccoli and the cauliflower heads on a fine mandoline or with a sharp knife, but stop when you reach the stalks. Discard the stalks and add the chopped broccoli and cauliflower to the rest of the vegetables.

Add the red onions, pour over the French dressing and lemon juice and leave to marinate for 30 minutes – no more or the vegetables will start to soften. Season to taste and serve.

SERVES 6 AS A LIGHT LUNCH

Griddled Butternut Squash, Goat's Cheese and Olive

This is a salad that plays on the contrast of colours and looks stunning layered up, rather than tumbled together. We make it with a soft, creamy goat's cheese.

1 large butternut squash, peeled, halved and deseeded

150ml extra virgin olive oil

150g soft goat's cheese

15g pumpkin seeds, lightly toasted in a dry pan

1 small bunch of fresh flat-leaf parsley, roughly chopped

fine sea salt and freshly ground black pepper

For the balsamic onions:

3 medium red onions

3 tablespoons extra virgin olive oil

1 tablespoon good balsamic vinegar

50g caster sugar

For the black olive purée:

100g pitted black Kalamata olives

1 tablespoon capers

1 clove of garlic

3 tablespoons extra virgin olive oil

For the garlic and chilli dressing:

200g crème fraîche

1 clove of garlic, crushed

½ a plump red chilli, deseeded and very finely chopped

freshly ground black pepper

Preheat the oven to 200°C/gas 6.

To make the balsamic onions, peel the onions and cut each one into 8 wedges, keeping the root end intact so the wedges don't fall apart. Scatter over a shallow ovenproof dish and add the olive oil, balsamic vinegar, sugar and 3 tablespoons of water. Toss lightly together, then cover the dish with foil, put into the oven and bake for 25 minutes. Remove the foil and cook for a further 5–10 minutes, or until the onions have softened and are lightly browned. Remove and leave to cool, but leave the oven on.

While the onions are baking, prepare the butternut squash. Cut the squash in half again, then cut into thick wedges (about 1.5cm). Heat a griddle pan preferably, or otherwise the grill, and cook the squash for 1–2 minutes, turning once, until lightly charred on both sides. Transfer

Griddled Butternut Squash continued

to a baking tray, brush generously with roughly a third of the oil and season. Put into the oven and cook for 10 minutes, then take out and turn the squash over. Brush with half the remaining oil and season generously. Return to the oven for a further 10 minutes, until softened and lightly browned, then remove and cool to room temperature.

To make the black olive purée, put all the ingredients into a blender and blitz to a coarse paste.

To make the dressing, put the crème fraîche into a bowl and stir in the garlic and chilli. Add enough cold water to make the dressing pourable (around 1–2 tablespoons) and season with a little ground black pepper.

To serve, arrange the butternut squash in a shallow serving dish in a single layer. Crumble the goat's cheese into chunky pieces and scatter them over the top, then drizzle the dressing over. Top with the balsamic onions. Dot with the olive purée and sprinkle with the toasted pumpkin seeds and parsley. Finally drizzle with the remaining oil and serve.

Variation: With Aubergine, Pomegranate, Feta and Pumpkin Seeds

Follow the recipe above, but omit the olive purée and substitute 3 medium aubergines for the squash. Trim the ends of the aubergines and cut into 1.5cm slices, then griddle or grill in the same way. Use 250g of drained feta instead of the goat's cheese, and sprinkle the seeds from a large ripe pomegranate (or 130g of seeds bought separately) over the salad along with the pumpkin seeds. Finish with fresh coriander, instead of parsley.

SERVES 6 AS A LIGHT LUNCH

Pickled Pear and Hazelnuts with Chickpeas, Quinoa and Daylesford Blue

In autumn and winter on the farm, the fresh pears that come in from the orchards go beautifully with our Daylesford Blue cheese from the creamery. You could use a creamy blue, such as Stichelton, Barkham Blue and of course Stilton instead, though ours will have a more crumbly texture. Pears and blue cheese are a classic combination, but here you need a relatively firm, yet ripe, flavoursome variety of pear, such as Williams or Conference, which can stand up to being cooked and pickled.

250g chickpeas

250g quinoa

120g toasted hazelnuts

100g sunflower seeds, toasted in a dry pan

5 tablespoons extra virgin olive oil

juice and zest of ½ a lemon

1 small bunch of fresh flat-leaf parsley, roughly chopped

150g blue cheese, such as Daylesford Blue, rind removed

sea salt and freshly ground black pepper

For the pickled pears:

2 firm pears, peeled, but with stalks left on

2 tablespoons sugar

100ml white wine vinegar

½ a red chilli, split and seeds removed

Soak the chickpeas in water overnight, then rinse and drain. Half fill a medium pan with water, add the chickpeas and bring to the boil, then turn the heat down to a simmer and cook for 40 minutes, or until tender. Drain and transfer to a large bowl to cool.

Half fill a medium pan with water and bring to the boil, then add the quinoa and turn the heat down to a simmer. Cook for 5 minutes, then remove from the heat and leave to stand for 5 minutes. Drain and transfer to a separate bowl to cool.

While the pulses are cooking and cooling, put the pears into a small pan with the sugar, vinegar, chilli and 800ml of water – the liquid should cover the pears. Bring to the boil, then turn down the heat and simmer gently for around 20 minutes, until the pears are tender. Leave to cool in their liquid, and once cool remove the stalks, cut in half, and remove the cores. Cut each half into 8 lengthways, to give 16 wedges in total.

Pickled Pear and Hazelnuts continued

Add the quinoa to the bowl of chickpeas, along with the pieces of poached pear, the nuts and seeds, olive oil, lemon juice and zest, and parsley, and toss lightly.

Taste and season as necessary, then tumble on to a large serving platter. Cut the cheese into rough cubes, dot over the salad, then fold them in very lightly, taking care not to break up the cheese too much, and serve.

SERVES 4 AS A LIGHT LUNCH

Lentils, Tomato, Daylesford Blue and Red Onion

A good all-year-round salad in which the nuttiness of the lentils balances with the slight saltiness of the blue cheese. Alternatives to Daylesford Blue include Stichelton, Barkham Blue, or possibly Stilton, though this has less creaminess. The salad needs to be served at room temperature so that the tomatoes can also show off their fullest flavour. Good varieties to use for this are Stupice early in the season, and Marmande later.

200g Puy lentils, rinsed and drained

2 large ripe tomatoes, finely chopped

1 medium red onion, finely chopped

150g blue cheese, rind removed

1 small handful of fresh dill, roughly chopped

1 small bunch of fresh chives, finely chopped

For the dressing:

1 tablespoon wholegrain mustard

1 teaspoon Dijon mustard

juice of ½ a lemon

1 clove of garlic, crushed

4 tablespoons olive oil

freshly ground black pepper

Half fill a medium pan with water and bring to the boil. Add the lentils to the pan, stir well and return to the boil. Cook for 18–20 minutes, or until just tender.

While the lentils are cooking, make the dressing. Put the mustards, lemon juice and garlic into a bowl and stir until combined. Whisk in the olive oil, a little at a time, until the dressing emulsifies. Season with black pepper.

Drain the lentils through a sieve and rinse under running water until cold. Drain well again and tip into a serving dish. Scatter the tomatoes and red onion over the top. Pour the dressing over, and toss well together.

To serve, cut the cheese into rough cubes and dot over the salad. Toss very lightly, so as not to break up the cheese too much, and scatter with the dill and chives.

SERVES 6

Curried Cauliflower, Red Pepper and Nigella Seeds

Cauliflower is a vegetable that lends itself to spice as well as sweet and sour flavours – in this dish the sweetness comes from raisins. Here, the key to cooking it is to use a big, wide pan to allow the cauliflower space to sauté and take on a little colour, before adding some water and letting it soften. If you overcrowd the pan, you will bring the temperature of the oil down and the cauliflower will steam, rather than fry.

50ml sunflower oil

2 teaspoons medium curry powder

2 teaspoons ground turmeric

1 teaspoon mustard seeds

1 teaspoon cumin seeds

1 medium white onion, finely sliced

1.4kg cauliflower, cut into small florets

2 large red peppers, deseeded and cut into thin strips

1 hot red chilli, deseeded and very finely chopped

1 tablespoon nigella (black onion) seeds, lightly toasted in a dry pan

1 medium bunch of coriander, leaves roughly chopped

85g raisins

½ a medium red onion, finely sliced

50g toasted flaked almonds

2 tablespoons olive oil

2 teaspoons lemon juice

2 teaspoons white wine vinegar

sea salt and freshly ground black pepper

Heat the oil in a large, wide, heavy-based pan and gently heat the curry powder, turmeric, mustard and cumin seeds for a few seconds, until the mustard seeds begin to pop. Add the onion and sauté until soft, stirring regularly.

Put the cauliflower, red peppers and chilli into the pan, making sure that the cauliflower is in a single layer, and cook over a medium heat for 5 minutes, until the cauliflower has lightly coloured, stirring regularly. Add 6 tablespoons of water and continue to cook for 5 more minutes, until the cauliflower has softened but still has a crunch to it (there should be no liquid remaining in the pan).

Remove the spiced vegetables from the heat, transfer to a large serving bowl and leave to cool. Then add the nigella seeds and all the rest of the ingredients, tumbling the salad lightly together and seasoning to taste.

SERVES 6

Chestnut, Quinoa, Kale and Broccoli

A feel-good salad for the winter, full of greens and goodness, that we put on the menu in the lead up to Christmas, when the first chestnuts arrive in the kitchen. The salad can be made with florets of green broccoli until the first purple sprouting broccoli appears from around the end of February right through to early May. The easiest thing is to use vacuum-packed chestnuts, but if you want to roast your own, preheat the oven to 200°C/gas 6. Cut a cross through the shell on the top of each nut, put them into a roasting tin, and let them roast in the oven for about 30 minutes, or until the shells have opened out and the nuts inside are tender. Let them cool, then peel away both the outer shells and the inner bitter skins.

The key to this salad is to blanch the broccoli and kale very, very briefly, to retain their crunch, colour and nutrients.

200g jarred, vacuum-packed or freshly roasted and peeled chestnuts

500g quinoa

250g purple sprouting broccoli tops

200g young curly kale leaves (rough stalks removed), shredded quite thickly

4 tablespoons hazelnut oil

4 tablespoons extra virgin olive oil

2 cloves of garlic, crushed

1 long red chilli, deseeded and finely chopped

zest and juice of 1 large orange

sea salt and freshly ground black pepper

If using vacuum-packed or jarred chestnuts, preheat the oven to 200°C/gas 6.

Spread the nuts out evenly on a baking tray and bake for 10 minutes, then remove and leave to cool.

Half fill a medium pan with water, add ½ teaspoon of sea salt and bring to the boil, then add the quinoa and turn the heat down to a simmer. Cook for 5 minutes, then remove from the heat and leave to stand for 5 minutes. Drain and transfer to a separate bowl to cool.

Half fill a large saucepan with water and bring to the boil. Put in the broccoli and return quickly to the boil, then immediately lift out with a slotted spoon (leaving the water in the pan). Transfer the broccoli to a colander and drain under plenty of running water until completely cold. Drain well and tip into the bowl with the quinoa.

Return the water to the boil, add the kale, bring back to the boil and cook for just 5 seconds. Drain in a colander under running water until completely cold. Drain well again and add to the broccoli and quinoa.

In a small bowl, mix the oils, garlic, chilli and orange juice, add a good pinch of salt and a couple of twists of freshly ground black pepper, then pour over the vegetables and quinoa, sprinkle with the orange zest, toss well together and serve.

Purple Sprouting Broccoli, Spelt, Crispy Garlic and Toasted Almonds

This is all about celebrating the arrival of purple sprouting broccoli around the end of February. The spelt, garlic crisps and almonds add a little crunch; crisping the garlic also gives bursts of flavour, which complements the broccoli, rather than overpowering it.

125g pearled spelt

800g purple sprouting broccoli (about 500g after stalks removed)

150ml olive oil

4 cloves of garlic, very finely sliced

20g toasted flaked almonds

zest and juice of 1 lemon

1 plump red chilli, deseeded and finely chopped

sea salt and freshly ground black pepper

Half fill a medium saucepan with water, add ½ teaspoon of salt, and bring to the boil. Stir in the spelt and return to a simmer. Cook for about 20–30 minutes, or until the spelt is just tender, stirring occasionally, then drain and rinse under running water until cold. Drain well and tip into a large bowl.

While the spelt is cooking, half fill a large saucepan with water and bring to the boil. Add the broccoli and bring back to the boil, then take off the heat immediately, so that the broccoli stays crisp. Drain through a colander under running water until completely cold. Leave to drain for 10 minutes, then tip into the bowl with the spelt.

Heat 100ml of the oil in a small frying pan and fry the garlic slices gently for 2–3 minutes until evenly golden and crisp, taking care not to let them burn or they will taste bitter. Remove from the heat, lift out with a slotted spoon and drain on kitchen paper.

Add the almonds, lemon zest and juice and chilli to the bowl of spelt and broccoli, along with the rest of the olive oil, a good pinch of salt and some freshly ground black pepper. Toss well, tip into a wide serving bowl or platter, sprinkle with the reserved crispy garlic slivers and serve.

SERVES 6

Raw Beetroot, Kidney Beans and Mustard Leaf with Horseradish Dressing

A wintry, quite hearty salad, making good use of the later, bigger beetroot and peppery winter salad leaves. If you can't find mustard leaves, use baby spinach instead.

Although you could use tinned beans, it is always preferable to cook your own. If you want to do this, soak 200g dried kidney beans in a large bowl and cover with cold water. Leave in a cool place for at least 8 hours or overnight. The next day, rinse the beans thoroughly. Half fill a large saucepan with cold water and add the beans, ensuring they are fully covered. Bring to the boil over a high heat for 10 minutes, then turn the heat down and cook very gently for 45–60 minutes, until tender – you really want the beans to be nice and soft for this; to test, take out one of the beans and press it with your thumb – it should give easily. Rinse in a colander under running water until cold, and drain before using.

6 small handfuls of mustard leaves or baby spinach leaves

300g raw beetroot

400g cooked red kidney beans, drained and rinsed

1 small bunch of fresh flat-leaf parsley, roughly chopped

For the horseradish dressing:

140g mayonnaise

100g hot horseradish sauce

125g natural yoghurt

sea salt and freshly ground black pepper

To make the dressing, mix the mayonnaise with the horseradish and yoghurt in a large bowl and season to taste.

Cut the larger leaves into thin strips. You can do this quickly by piling 4 or 5 leaves on top of each other and slicing them together. Trim and peel the beetroot, then grate coarsely, either by hand or using a food processor. Bearing in mind that the beetroot juice will stain, you may want to use latex gloves to protect your hands.

Add the grated beetroot to the bowl of dressing, along with the kidney beans, leaves and chopped parsley. Mix thoroughly, then taste and adjust the seasoning if necessary before serving.

Mushroom, Celeriac, Truffle Honey and Toasted Pine Nuts

Ceps and celeriac are around at the same time in the autumn, and for this salad, they are just sliced and marinated, then finished off with winter micro-cress or watercress.

You can usually buy truffle honey in good delis, but if you can't find it, mix 7 tablespoons of clear honey with 1 tablespoon of black truffle paste. And if you are able to treat yourself to some white truffle, then a little, grated over the salad at the end, will make it extra special.

Soak the pine nuts for the purée overnight or for around 4–6 hours, as this will soften them and result in a much smoother texture.

400g cep mushrooms, cleaned

400g chestnut mushrooms, cleaned

6 tablespoons white wine vinegar

5 tablespoons extra virgin olive oil

2 tablespoons chopped fresh flat-leaf parsley

1 large head of celeriac

3 tablespoons pine nuts, toasted in a dry pan

1 good handful of micro-cress or watercress

8 tablespoons truffle honey (see introduction, above)

5g white truffle (optional), to finish

For the pine nut purée:

200g pine nuts

4 tablespoons extra virgin olive oil

juice of ½ a lemon

freshly ground black pepper

Soak the pine nuts for the purée in water overnight, then drain and put into a blender with the olive oil, lemon juice and 5 tablespoons of water. Season with black pepper and blend, adding a little more water if necessary until you have a smooth purée consistency. Transfer to a bowl and put into the fridge while you prepare the rest of the salad.

Thinly slice the mushrooms on a mandoline or with a sharp knife, and put them into a bowl with 2 tablespoons of the white wine vinegar, 4 tablespoons of the olive oil and the parsley. Season and leave for about 20 minutes.

Peel the celeriac and cut it in half. To get a good mixture of shapes and textures, take one half and cut it into strips about 6cm long and 5mm wide. Slice the other half very thinly, using a mandoline or a sharp knife, then cut the slices into very thin spaghetti-like strips. Put all the celeriac into a separate bowl, mix with the remaining vinegar and olive oil, and season.

To serve, divide the pine nut purée between four plates. Lift the mushrooms and celeriac out of their marinades, and scatter over the top, followed by the toasted pine nuts and the cress. Drizzle with the truffle honey and finish with a fine shaving of fresh white truffle, if using.

Wild Rice, Red Cabbage, Apple and Toasted Cobnuts

SERVES 4

From mid to late August through to September, we can gather cobnuts, the cultivated version of wild hazelnuts, from the trees around the farm – you have to be quick, though, to beat the squirrels. In season you should be able to find the nuts in shops, greengrocers and markets. If you can't find cobnuts, you can use thinly sliced hazelnuts (though these have a slightly more intense flavour), or whole almonds, again thinly sliced.

Cobnuts and apples are ready for harvesting at the same time – and we make this salad with many different apple varieties throughout the season, including heritage ones, such as Blenheim Gold, a popular, old, local variety. You need apples with some sweetness and preferably some good streaks of red in the skin, to give a nice colour to the salad, so of the more readily available apple varieties, Jonagold is a good one to choose.

200g wild rice, rinsed in cold water

150g red cabbage

½ a medium red onion, finely sliced

2 red-tinged eating apples, such as Jonagold

25g watercress

5 tablespoons finely chopped fresh flat-leaf parsley

40g thinly sliced cobnuts, lightly toasted in a dry pan

juice of 1 lemon

2 tablespoons extra virgin olive oil

1½ tablespoons good balsamic vinegar

1 teaspoon fine sea salt

a few twists of freshly ground black pepper

Half fill a medium pan with water, bring to the boil, then add the wild rice and stir a couple of times. Bring up to a simmer and cook for 35–40 minutes, or until the rice is tender. Drain in a sieve under running water until the rice is cold, then drain well again, and tip into a mixing bowl.

Trim the cabbage of any damaged outer leaves, cut in half and remove the central core, then slice the leaves very finely. Add to the rice, along with the red onion. Quarter and core the apples, grate coarsely then add to the bowl, with the rest of the ingredients, toss gently and serve.

SERVES 6 AS A LIGHT LUNCH

Chicken Caesar

This is another recipe that has become a staple on the salad counter at our farmshops. It is one of those recipes where there is some debate over the 'correct' ingredients but we've gone with what makes use of ingredients we can source on the farm. The original recipe is believed to have been dreamed up by American restaurateur Caesar Cardini in 1924, and was quite sparing: just crunchy lettuce, dressing, croutons and grated Parmesan; but over the centuries all kinds of variations have found their way onto menus. Whatever ingredients are used, two things can let a Caesar salad down: the dressing can be just a creamy medium with no real flavour and the croutons can be big and either far too hard and crunchy or so soft they soak up too much oil and become greasy.

Our chefs put capers in the dressing, which gives the dressing a lift, and, instead of croutons, we make thin, baked toasts using our seven seeds sourdough (see page 299). They give the essential crunch to the salad, but they are lighter and not greasy.

3 boneless, skinless chicken breasts

3 tablespoons olive oil

4 medium eggs

6 rashers of rindless smoked streaky bacon

6 very thin slices of multi-seed bread

6 little gem lettuces, separated into leaves

75g Parmesan cheese, shaved thinly

sea salt and freshly ground black pepper

For the Caesar dressing:

40g Parmesan cheese, finely grated

1 tablespoon capers in vinegar, drained, plus 2 teaspoons of their vinegar

2 teaspoons fresh lemon juice

1 plump clove of garlic, roughly chopped

freshly ground black pepper

200g mayonnaise

Preheat the oven to 200°C/gas 6.

Place the chicken on a baking tray and season. Rub with 2 tablespoons of the oil and then roast for 15 minutes, until very lightly browned and cooked throughout (the juices should run clear if the chicken is pierced with the tip of a sharp knife). Transfer to a plate and leave to cool, but leave the oven on, turning down the heat to 180°C/gas 4.

While the chicken is cooking, half fill a medium pan with water and bring to the boil. Gently lower the eggs into the water and return it to the boil. Cook for 7 minutes, drain the eggs under running water until cold, then peel.

Put a large non-stick frying pan over a medium heat and add the bacon. Cook for about 3 minutes on each side until very crisp, then remove and drain on kitchen paper.

Arrange the bread on a baking tray, drizzle with the rest of the olive oil and season with a pinch of salt and freshly ground black pepper. Put into the oven and bake for 6–8 minutes, until very crisp and golden brown. Remove and leave to cool while you make the dressing.

For the dressing, put all the ingredients, except for the mayonnaise, into a food processor, season with ground black pepper, and blitz until as smooth as possible. You may need to remove the lid and push the mixture down with a rubber spatula once or twice. Transfer to a small bowl and stir in the mayonnaise.

To assemble, cut the chicken breasts into small chunks. Quarter the eggs. Scatter the lettuce over a serving dish or platter and arrange the chicken on top, then drizzle with the Caesar dressing. Break the crispy bacon and toasted bread into rough pieces and scatter over, finish with the eggs and sprinkle with the Parmesan.

VEGETABLES

Notes on vegetables

Vegetables from the market garden are at the heart of so much of the cooking we do at the farm, and we are always spoilt for choice. What we grow informs the menus at the cafés – our chefs change and devise them according to the produce that is available in the garden – and for our cookery school it's the first port of call when showing visitors where the ingredients they are cooking with come from. Although we are a mixed farm and we raise livestock, we believe that vegetables need to be at the centre of the plate and aside from the flavours, nutrition and texture that vegetables bring, above all for me, the garden is so integral to the work we do because it demonstrates why we believe eating seasonally is so important.

We live in a world where many of us have access to most types of produce all the year round, but to me it makes no sense to eat food that's been flown half way around the world to reach us. For me, eating food that has been grown locally, in harmony with nature's natural cycles, is the right way to eat. Nutritionally it is better for our bodies; it is helping us to protect and secure the long-term health of our planet and, I believe, it is infinitely better in flavour. How can a tomato bought in January – an insipid, watery fruit that has travelled or been forced to grow without sunlight – taste as fresh or as vibrant as a pea that's been picked on a warm sunny spring day? Even putting aside the environmental and sustainability arguments, food that grows according to its seasonal climate and has not had to sustain months of storage and travel simply tastes better.

Some of the recipes in this chapter can be served as main courses, such as the woodland mushroom shepherd's pie on page 128, the spiced pumpkin, butter bean and spinach casserole on page 134 and the beetroot, swede and potato bake on page 131, while others will need something alongside them to make a full meal – a grain or some sourdough bread, just to fill them out. Most are very simple so making two or three and combining them will give you a meal that's colourful and full of a diverse range of vitamins and nutrients.

SERVES 4 AS AN ACCOMPANIMENT

Crushed New Potatoes with Olives, Capers and Herbs

This combination goes particularly well with fish and is lovely when the new potatoes are in season, around May for Jersey Royals and June for other varieties. Of course you could still make this at other times of the year with maincrop potatoes – peel them first, though.

650g new potatoes, such as Jersey Royals, skin on

50g pitted Kalamata black olives, drained and chopped

4 tablespoons capers, drained

¼ of a small fennel bulb, trimmed and very finely grated

4 tablespoons finely chopped fresh parsley

5 tablespoons extra virgin olive oil

juice of 1 lemon

sea salt and freshly ground black pepper

Put the potatoes into a pan of cold, lightly salted water and bring to the boil, then turn down the heat and simmer until tender. Drain, then return the potatoes to the pan, crushing them slightly with a fork, and add the rest of the ingredients. Place over a low heat and warm through for 3–4 minutes, stirring regularly until hot throughout.

SERVES 4 AS AN ACCOMPANIMENT

Smashed Broad Beans, Peas and Mint

A celebration of the height of summer: fresh vibrant flavours.

- 250g fresh peas
- 250g podded broad beans
- 3 tablespoons chopped fresh mint
- 3 tablespoons finely chopped fresh chives
- 2 handfuls of pea shoots
- 50g butter
- sea salt and freshly ground black pepper

Have ready a bowl of iced water.

Half fill a large saucepan with water and bring to the boil. Add the peas first if fresh and cook for about 5 minutes, adding the beans for the last 30 seconds. If the peas are frozen, you can put them in together and bring back to the boil, then take off the heat after 30 seconds. Drain in a colander, then transfer to the bowl of iced water and leave for 5 minutes. Drain well again and slip off the skins from the broad beans.

Transfer the peas and beans to a bowl and crush with the back of a fork. Stir in the herbs and pea shoots and stir well.

Melt the butter in a large non-stick pan. Add the crushed bean and pea mixture and heat gently, stirring regularly, until warmed through. Season and serve.

SERVES 6 AS AN ACCOMPANIMENT

Wilted Spinach with Toasted Pine Nuts, Sultanas and Lemon Zest

This has a Mediterranean feel and goes well with roast chicken or slowly roasted lamb shoulder, along with the broad bean, bulgar wheat and herb salad (see page 73).

50g pine nuts
40g butter
1 medium onion, thinly sliced
2 cloves of garlic, finely chopped
50g sultanas
4 tablespoons white wine
360g baby spinach, washed and well dried
zest of 1 lemon
sea salt and freshly ground black pepper

Lightly toast the pine nuts in a dry pan over a low heat until just golden – take care not to let them burn.

In a medium-sized pan, melt the butter over a low heat, then add the onion and garlic and continue to cook for about 5 minutes, until the onion is soft, but not coloured.

Add the sultanas and white wine and cook for a further 5 minutes, until all the liquid has gone. Add the toasted pine nuts and spinach and cook for a further minute, until the spinach has wilted. Season and add the lemon zest. Stir well and serve.

SERVES 6 AS AN ACCOMPANIMENT

Green Beans with Almonds, Parsley and Garlic Butter

The almonds add a lovely crunch to this very simple but flavoursome side dish.

500g green beans, tops removed
80g butter
60g flaked almonds
4 cloves of garlic, finely chopped
2 tablespoons roughly chopped fresh flat-leaf parsley
juice of ½ a lemon
sea salt and freshly ground black pepper

Half fill a large pan with lightly salted water and bring to a rapid boil. Drop in the prepared green beans and cook over a high heat for about 8 minutes, until the beans are fully cooked but still have a little bite to them, then drain.

Heat the butter in a frying pan until it starts to foam, then cook for a couple more minutes until it becomes golden brown and takes on a nutty flavour – but take care not to let it burn.

Add the flaked almonds and continue to cook, stirring constantly, until the almonds are golden brown and toasted. Add the drained beans, garlic and parsley and season with salt and pepper.

Add the lemon juice, toss well and serve immediately.

Potato Wedges with Garlic and Rosemary Butter

These wedges are the Daylesford alternative to chips, according to our chefs. They say that the key is to do as you would when boiling potatoes before roasting them – slightly roughen and break them up around the edges, so that when they go into the oven, you get edges and corners that turn crispy and crunchy.

- 1.2kg baking potatoes, such as Sante, Cara, or the red Romano, well scrubbed
- 1 tablespoon olive oil
- 25g butter, at room temperature
- 1 clove of garlic, finely crushed
- 1 tablespoon chopped fresh rosemary leaves
- 1 tablespoon finely chopped fresh flat-leaf parsley
- sea salt and freshly ground black pepper

Preheat the oven to 200°C/gas 6.

Cut the potatoes into wedges, put them into a pan of cold, lightly salted water and bring to the boil. Turn down the heat and simmer for about 5 minutes, or until tender, but not breaking apart. Drain in a colander and leave to steam for 5 minutes, shaking the colander gently to roughen the exposed surface of the potatoes.

Pat any remaining moisture from the potatoes with some kitchen paper and arrange them over a baking tray in a single layer. Drizzle with the olive oil and season. Toss lightly, put into the oven and bake for 15–20 minutes, turning once, until golden brown.

Meanwhile, mix the butter with the garlic and rosemary until thoroughly combined.

Once the potatoes are golden, dot them with pieces of the garlic butter and return them to the oven until the butter has melted. Remove and season again, if necessary, then serve immediately, scattered with parsley.

SERVES 4 AS AN ACCOMPANIMENT

Crushed, Buttered Root Vegetables and Cabbage

This combination works really well alongside winter stews and casseroles that have big robust flavours. The crushed vegetables should look quite rustic and the key is to chop them fairly small, not in great chunks, so that you can just crush them with a fork and retain a good texture and a little of their shape here and there – then the fine shreds of cabbage are just mixed through.

2 medium carrots, roughly chopped

½ a medium swede, roughly chopped

½ a medium Savoy or white cabbage, tough core removed, finely shredded

40g butter

sea salt and freshly ground black pepper

Put the carrots and swede into a pan of cold, lightly salted water and bring to the boil, then turn down the heat and simmer for 25–30 minutes, until tender. Lift out with a slotted spoon (leaving the water in the pan), transfer to a bowl and leave to cool.

Return the pan of vegetable water to the heat and put in the cabbage. Bring back to the boil and cook for 1 minute, then drain well in a sieve and pat dry.

When ready to serve, melt the butter in a large non-stick pan over a low heat. Add the carrot and swede and crush lightly with the back of a fork until they are roughly mashed, but still have some texture. Stir in the cabbage, season and cook for 3–5 minutes, stirring regularly, until hot throughout. Serve immediately.

SERVES 6

Woodland Mushroom Shepherd's Pie

This is a wonderful earthy, hearty vegetarian alternative to shepherd's pie for the autumn. It's best to cook the mushrooms in batches, so that they sauté properly and become nice and golden. If you pile too many into the pan at the same time, you will reduce the heat and they will steam, rather than fry. If you can't find enough of the mushroom varieties suggested, you can add some chestnut mushrooms to make up the weight. When you chop the vegetables they should be small – about 1cm.

500g mixed wild mushrooms, such as portabello, ceps, chanterelles and pied bleu

75g butter

4 tablespoons olive oil

3 medium white onions, finely chopped

2 sticks of celery, chopped

1 carrot, chopped

¼ of a swede, chopped

6 cloves of garlic, finely chopped

100ml sherry

200ml white wine

300ml double cream

1 teaspoon Dijon mustard

1 tablespoon Worcestershire sauce

a pinch of cayenne pepper

2 tablespoons chopped fresh flat-leaf parsley leaves

3 tablespoons Parmesan cheese, finely grated

For the mash:

5 large potatoes, peeled and quartered

70g butter

50ml milk

sea salt and freshly ground black pepper

First make the mash. Put the potatoes into a medium pan, cover with cold, lightly salted water and bring to the boil, then turn down the heat and simmer for around 20 minutes, until the potatoes are cooked through and easily fall away if pierced with the tip of a sharp knife. Drain in a colander and leave to steam briefly, then return them to the pan and mash well (alternatively put them through a potato ricer, then return them to the pan). Add the butter and milk and season. Leave the pan at the side of the stove while you prepare the mushrooms.

Divide the mushrooms into 4 equal mounds, as you want to fry them in batches. Also cut the butter into 5 equal pieces – set one aside to cook the onions and other vegetables in a moment. Heat one of the remaining 4 pieces of butter with a tablespoon of oil in a large non-stick frying pan. Add the first mound of mushrooms and fry for a couple of minutes, until golden brown, then remove. Put in the next piece of butter and another tablespoon of oil and when the butter has melted fry the next mound of mushrooms. Repeat twice more.

In a casserole (one that has a lid), melt the reserved piece of butter and add the onions, celery, carrot, swede and garlic. Cook without a lid over a medium heat for 10 minutes, until the vegetables are starting to colour slightly and become soft.

Add the cooked mushrooms, together with the sherry and wine, and bring to the boil. Simmer for around 10 minutes, until nearly all the liquid has gone.

Add the cream, mustard, Worcestershire sauce, cayenne and parsley, and season. Bring back to the boil, then take off the heat straightaway, taste and season again if necessary.

Preheat the oven to 180°C/gas 4.

Spoon the mixture into a deep pie dish and either spoon or pipe the reserved mashed potatoes over the top. Sprinkle with the grated Parmesan.

Put into the oven for about 45 minutes, until golden brown on top and hot all the way through.

SERVES 6–8

Beetroot, Swede and Potato Bake

Beetroot and swede add extra flavours and colour to what is essentially a potato dauphinois. This is a wonderfully comforting autumnal dish that is great served with a crunchy green vegetable to add freshness and complement its richness.

450ml double cream	2 large baking potatoes, peeled
450ml milk	3 large beetroot, peeled
2 cloves of garlic, thinly sliced	2 medium swede, peeled
1 large sprig of fresh thyme, leaves only	sea salt and freshly ground black pepper

Preheat the oven to 170°C/gas 3.

Put the cream, milk, garlic and thyme leaves into a medium pan and bring to a simmer, then immediately remove from the heat, season and leave to stand, so that the flavours infuse.

With a mandoline or sharp knife, slice each of the vegetables very finely (about 2mm), keeping each type separate.

Place a thin layer of potato in a deep baking dish, season, then spread with 3 tablespoons of the cream mixture. Next put in a thin layer of beetroot, season and add 3 more tablespoons cream. Follow with a thin layer of swede, season and spoon over 3 more tablespoons of cream. Continue until all the vegetables are used up, but keep back enough potato to make sure that you finish with this. Season again, then pour any remaining creamy liquid over the top.

Push the vegetables down gently with your fingers to ensure that all the layers are fully immersed in the liquid. Cover with a sheet of greaseproof paper and then some foil. With the tip of a sharp knife, prick a few holes through both the foil and the greaseproof paper, to allow some steam to escape during cooking. Place the baking dish on a baking tray, as the cream will probably bubble over. Put into the oven for 1 hour, then remove the foil and continue to bake for 10 more minutes, until golden brown on top. Remove from the oven, leave to settle for 5 minutes, then serve.

SERVES 4

Baked Leeks with Cider

Leeks have a natural sweetness which is brought out when you cook them slowly in butter; the cider complements this perfectly.

4 leeks

40g butter

1 clove of garlic, thinly sliced

500ml dry cider

2 tablespoons cider vinegar

400ml good vegetable stock

1 sprig of fresh thyme, leaves only

200ml double cream

sea salt and freshly ground black pepper

For the Parmesan and herb crumbs:

55g breadcrumbs

2 tablespoons fresh flat-leaf parsley leaves

1 clove of garlic, finely chopped

½ teaspoon sea salt

freshly ground black pepper

2 tablespoons grated Parmesan cheese

Preheat the oven to 180°C/gas 4.

Trim the leeks, removing and discarding the outer layer. Cut off the green part of the leeks about 2.5cm into the white part, and then cut the green (and little bit of white) into 1cm slices. Wash well and drain in a colander. Cut the remaining white part of the leeks into 4 equal lengths, leaving these as whole cylinders. By cutting off the green parts quite low down, into the white area, you will have taken away the bits that need a good wash to remove soil that is trapped inside, so the tight white cylinders need only a rinse on the outside.

In a medium pan melt the butter and add the chopped green of the leek, together with the garlic. Allow to soften for a few minutes over a medium heat, then add the cider and cider vinegar and bring to the boil. Cook for about 10 minutes on a high heat, until all the liquid has gone.

Add the stock and thyme leaves, then bring back to the boil and cook until only a quarter of the liquid remains. Add the cream and again bring back to the boil, then remove from the heat straightaway. Taste and season.

Lift the green of the leek out of the liquid and place in the bottom of a baking dish. Arrange the raw white cylinders of leek on top in rows. Spoon over the creamy liquid and season with a little more salt and pepper. Cover with foil and place in the oven for 30 minutes.

Whilst the leeks are baking, put the breadcrumbs, parsley, garlic, salt and pepper into a blender or food processor and whizz until you have fine green crumbs, then add the grated Parmesan and continue to blend for 30 seconds more.

Remove the leeks from the oven, take off the foil and sprinkle the herby breadcrumbs over the top. Put back into the oven for a further 10 minutes to brown slightly, then serve.

SERVES 6

Spiced Pumpkin, Butter Bean and Spinach Casserole

Casseroles are a staple on my table throughout the autumn and winter. At this time of year I crave warming, nourishing recipes and the soft texture of a casserole feels comforting but also bolstering.

2 tablespoons olive oil

400g pumpkin, peeled, deseeded and cut into chunks (about 3cm)

1 small onion, finely sliced

1 stick of celery, chopped

1 medium carrot, chopped

2 cloves of garlic, finely chopped

2.5cm piece of fresh root ginger, finely chopped

1 teaspoon ground cumin

1 teaspoon coriander seeds

½ teaspoon ground turmeric

½ a long red chilli, finely chopped

200g tomatoes, chopped

3 tablespoons honey

200g jarred or tinned butter beans, drained and rinsed

50g Puy lentils, rinsed

50g young spinach leaves, sliced

1 small bunch of fresh basil, finely chopped

2 tablespoons finely chopped fresh coriander

sea salt and freshly ground black pepper

Heat the oil in a large non-stick pan or flame-proof casserole. Add the pumpkin, onion, celery, carrot, garlic and ginger, and cook, stirring constantly, over a medium heat for 5 minutes, or until the vegetables are beginning to soften and are lightly browned. Add the spices and chilli and cook for 3 minutes more, again stirring constantly, and being careful not to let them burn.

Next add the tomatoes, honey, beans, lentils, 1 litre of water and 1½ teaspoons of sea salt and bring to a simmer. Cook for 25–30 minutes, stirring regularly, until the liquid is well reduced and the pumpkin has broken down and blended into it, so that the dish looks like a curry.

Stir in the spinach, basil and coriander and cook for 1–2 minutes more. Taste, adjust the seasoning and serve.

Five Risotti

The rice in a great risotto should retain a little bite to it, but at the same time the risotto should be quite loose and creamy – not soupy. If you tap the underneath of the plate or dish in which you are serving it, the risotto should flatten out. Usually the creaminess is achieved by beating in plenty of butter, or a combination of butter and extra Parmesan before serving. Ours differs a little in that our chefs add onion purée, which gives the risotto velvety texture without the addition of extra richness. The purée also makes a nice little sauce and you can make up a big batch of it and keep it portioned in the freezer, then defrost it just before you want to use it.

In a risotto, good stock is as vital as it is in soups, so either buy a good vegetable one or, preferably, make your own (see page 372).

MAKES ENOUGH FOR 9–10 RISOTTI

Onion Purée

Preheat the oven to 150°C/gas 2.

Put 1kg thinly sliced onions and 2 cloves of finely chopped garlic into a medium casserole (one that has a lid) with 300g butter, 2 tablespoons chopped fresh oregano, a sprig of fresh thyme (leaves only) and 500ml of water. Season with a little salt and pepper.

Cover and put into the oven for 1½ hours, or until the onions have completely softened but there is still a little liquid in the bottom of the casserole. Take out of the oven, leave to cool a little then blend to a smooth purée. Divide into portions and freeze until ready to defrost and use in risotto – or to accompany roasted meat.

Broad Bean, Pea and Watercress Risotto

SERVES 4

160g peas, fresh or frozen

160g podded broad beans

45g butter

1 tablespoon olive oil

1 small onion, finely chopped

1 clove of garlic, crushed to a paste

150g carnaroli or arborio risotto rice

50ml white wine

700ml good vegetable stock, hot (see page 372)

150g onion purée (see left)

1 tablespoon chopped fresh mint

4 tablespoons watercress and pumpkin seed pesto (see page 339)

60g Parmesan cheese, finely grated, plus 30g, shaved with a vegetable peeler, to finish

1 handful of watercress, roughly chopped

2 tablespoons extra virgin olive oil

sea salt and freshly ground black pepper

Half fill a large saucepan with water and bring to the boil. Add the peas first, if fresh, and cook for about 5 minutes, adding the broad beans for the last 30 seconds. If the peas are frozen, you can put the peas and beans in together and bring back to the boil, then take off the heat after 30 seconds. Drain in a colander under running water, then slip off the skins from the broad beans. Keep to one side.

Melt 15g of the butter with the oil in a heavy wide-based pan. Add the onion and garlic and soften gently without colouring, then add the rice and cook for 5 minutes, stirring constantly, over a low heat.

Add the white wine, turn up the heat a little and let the liquid reduce until it has almost gone, then add the hot stock a ladleful at a time, stirring continually and only adding the next ladleful when the previous one has been absorbed. Keep going until all the stock has gone and you have a creamy, quite loose consistency – the rice should still have a little bite to it.

Stir in the onion purée, the rest of the butter and the reserved broad beans and peas, and let the vegetables warm through. Stir in the mint, pesto and grated Parmesan, and taste and season as necessary.

Divide the risotto between four warmed bowls or deep plates, and garnish with the Parmesan shavings and chopped watercress. Drizzle with the extra virgin olive oil and serve immediately.

SERVES 4

Jerusalem Artichoke Risotto with Garlic and Almond Breadcrumbs

45g butter

1 tablespoon olive oil

1 small onion, finely chopped

1 clove of garlic, crushed to a paste

150g carnaroli or arborio risotto rice

50ml white wine

700ml good vegetable stock, hot (see page 372)

60g Parmesan cheese, grated

1 handful of micro- or baby salad leaves, to finish

2 teaspoons truffle oil, to finish

sea salt and freshly ground black pepper

For the artichoke purée:

320g Jerusalem artichokes, peeled and roughly chopped

250ml milk

50g butter

1 tablespoon lemon juice

For the breadcrumbs:

2 tablespoons olive oil

30g breadcrumbs

1 tablespoon chopped fresh rosemary leaves

1 teaspoon chopped fresh thyme leaves

½ a clove of garlic, crushed

15g flaked almonds

To make the artichoke purée, put the artichokes and milk into a pan with half the butter and 400ml of water. Bring to the boil, then reduce the heat and simmer until the artichokes are soft. Lift them out and transfer to a blender, adding just enough of the cooking liquid (about 4 tablespoons) to form a purée. Add the rest of the butter and the lemon juice and blend again. Taste and season with salt as necessary, then keep to one side.

For the risotto, melt 15g of the butter with the oil in a heavy wide-based pan. Add the onion and garlic and soften gently without colouring. Add the rice and cook for 5 minutes, stirring constantly, over a low heat.

Add the white wine, turn up the heat a little and let the liquid reduce until it has almost gone, then add the hot stock a ladleful at a time, stirring continually and only adding the next ladleful when the previous one has been absorbed. Keep going until all the stock has gone and you have a creamy, quite loose consistency – the rice should still have a little bite to it.

Jerusalem Artichoke Risotto continued

Add the artichoke purée, the Parmesan and the rest of the butter, stir and take off the heat. Taste and season as necessary.

For the breadcrumbs, heat the oil in a pan and quickly fry the breadcrumbs with the herbs and garlic. Drain on kitchen paper, then combine with the almonds.

Divide the risotto between four warmed wide bowls or deep plates, and garnish with the garlic breadcrumbs and the micro- or baby leaves. Drizzle with the truffle oil, finish with some freshly ground black pepper and serve immediately.

Leek and Wild Garlic Pesto Risotto

SERVES 4

15g butter

1 tablespoon olive oil

1 small onion, finely chopped

1 clove of garlic, crushed to a paste

150g carnaroli or arborio risotto rice

50ml white wine

700ml good vegetable stock, hot (see page 372)

2 leeks, finely chopped

150g onion purée (see page 136)

50g Parmesan cheese, grated, plus an extra 30g, shaved with a vegetable peeler, to finish

60ml wild garlic and pumpkin seed pesto (see page 339)

1 small bunch of fresh chives, chopped

1 handful of micro- or baby salad leaves

2 tablespoons extra virgin olive oil

sea salt and freshly ground black pepper

Melt 15g of the butter with the oil in a heavy wide-based pan. Add the onion and garlic and soften gently without colouring.

Add the rice and cook for 5 minutes, stirring constantly, over a low heat.

Add the white wine, turn up the heat a little and let the liquid reduce until it has almost gone, then add the hot stock a ladleful at a time, stirring continually and only adding the next ladleful when the previous one has been absorbed. Keep going until all the stock has gone and you have a creamy, quite loose consistency – the rice should still have a little bite to it.

Stir in the leeks and heat through, then stir in the onion purée and the grated Parmesan and after 30 seconds add the wild garlic pesto and the chives. Taste and season accordingly.

Divide the risotto between four warmed wide bowls or deep plates, and garnish with the Parmesan shavings and the micro- or baby leaves. Drizzle with the olive oil and serve immediately.

SERVES 4

Spelt, Garden Vegetable and Herb Risotto

120g podded broad beans

100g pearled spelt

75g butter

1 small summer squash, such as Crown Prince, peeled, deseeded and chopped (about 1cm)

2 leeks, finely chopped

1 tablespoon olive oil

1 small onion, finely chopped

1 clove of garlic, crushed to a paste

150g carnaroli or arborio risotto rice

50ml white wine

700ml good vegetable stock, hot (see page 372)

150g onion purée (see page 136)

2 courgettes, finely chopped

100g baby spinach

100g Parmesan cheese, finely grated

2 tablespoons chopped fresh herbs, such as parsley, sage and thyme

1 handful of micro- or baby salad leaves

sea salt and freshly ground black pepper

Half fill a large saucepan with water and boil. Add the beans, bring back to the boil, then take off the heat straightaway. Drain under running water, then slip the skins from the beans. Keep to one side.

Put the spelt into a pan of cold water and bring to the boil. Turn down the heat to a simmer for 20 minutes, until the grains are just cooked, then drain and set aside.

Meanwhile, heat 60g of the butter in a frying pan and sauté the squash until it starts to soften. Add the leeks and continue to cook until they are just beginning to colour, then remove from the heat.

Melt the rest of the butter with the oil in a heavy wide-based pan, then add the onion and garlic and cook gently without colouring. Add the rice and cook for 5 minutes, stirring constantly, over a low heat. Add the white wine, turn up the heat a little and let the liquid reduce until it has almost gone, then add the hot stock a ladleful at a time, stirring continually and only adding the next ladleful when the previous one has been absorbed. Keep going until all the stock has gone and you have a creamy, quite loose consistency – the rice should still have a little bite to it.

Stir in the onion purée, cooked spelt, courgettes, spinach, reserved broad beans and the sautéd vegetables and let them heat through. Add the Parmesan and herbs, taste and season accordingly. Divide the risotto between four warmed wide bowls or deep plates, garnish with the leaves and serve immediately.

SERVES 4

Pearl Barley, Asparagus and Pea Shoot Risotto

90g butter

1 medium onion, finely chopped

2 cloves of garlic, crushed

285g pearl barley

1.35 litres good vegetable stock, hot (see page 372)

150g slender asparagus, stalks thinly sliced, tips left whole

125g onion purée (see page 136)

2 teaspoons fresh lemon juice

60g young spinach (leaves only) and thinly sliced

4 spring onions, white parts only, thinly sliced

1 tablespoon finely chopped fresh mint leaves

75g Parmesan cheese, finely grated, plus an extra 65g, shaved with a vegetable peeler, to finish

1 small handful of pea shoots

2 teaspoons extra virgin olive oil

sea salt and freshly ground black pepper

Melt 50g of the butter in a large non-stick saucepan or casserole and gently cook the onion and garlic until softened, stirring occasionally.

Rinse the barley in a sieve under running water and add to the pan. Stir in 950ml of the stock and bring to the boil, then turn down the heat and simmer for 25–30 minutes, or until the barley is swollen and tender and most of the liquid has evaporated. Take off the heat.

Heat the remaining vegetable stock in a small pan and blanch the asparagus tips for 3 minutes, until just cooked, then lift out with a slotted spoon. Keep to one side and add the stock to the pan of barley. Bring this back to a gentle simmer, stirring regularly, then add the onion purée, lemon juice, spinach, spring onions, asparagus stalks, mint, grated Parmesan and the rest of the butter. Cook for 2 minutes, stirring constantly. Season with salt and pepper.

Divide the risotto between four warmed wide bowls or deep plates, and garnish with the reserved asparagus tips, Parmesan shavings and pea shoots. Drizzle with a little extra virgin olive oil, and serve.

The Creamery and Cheese Room

Making cheese by hand in the traditional way, as we do, can result in surprisingly different outcomes. It's a bit like cooks following the same recipe; ingredients behave differently depending on how you treat them. The protein and fat content in the milk can vary according to the season, a sudden change in the weather or because the cows have eaten a particular herb or a weed in the meadows. But that is what is fascinating and challenging about artisan cheese-making. Differences and nuances are all part of the attraction, whereas in industrial-scale cheese-making it is all about consistency: every batch must look and taste the same.

Some cheesemakers use an analysis of the milk to decide how to make each batch of cheese in advance, but our cheesemakers prefer to take it as it comes and judge it as they go along. Every style of cheese is made slightly differently, but once you add your culture to the milk and warm it up to start things off, the process can be quite quick or relatively slow, the cheesemaker is guided by the cheese and works with the milk's characteristics.

We develop and produce around nine or ten different cheeses at any one time, but the one I'm particularly fond of is our Single Gloucester. There are only a few farms that produce Single Gloucester as it now has Protected Designation of Origin status (PDO) and can only be made with the milk from a farm that has a registered herd of Old Gloucester cattle, a heritage breed that we introduced to Daylesford in 2006.

Single Gloucester dates back over two centuries. It was what was known as a kitchen cheese, as it was made with the partially skimmed milk left over from making Double Gloucester cheese and butter on small farms. Whereas Double Gloucester would be sold, the Single Gloucester was kept for the family and was rarely seen beyond a particular farm or village, let alone Gloucestershire. It was a cheese that had been almost lost and barely anyone remembers how the original was made, so the modern-day Single Gloucester has been reinvented and is made to a standardised recipe. As a result it is now acclaimed and sought after.

The two Gloucester cheeses are made quite differently, with different starters and maturing times. Single Gloucester is a light, fresh, buttery-tasting young cheese that has quite a bit of character – you can really taste the flavour of the milk in it – and is best eaten when it is around six to eight weeks old. If you keep it too much longer, it can dry out. Double Gloucester is much closer to a Cheddar, but with a more mellow flavour, and is traditionally coloured with annatto, a natural vegetable dye, to give it its orangey appearance. It is also matured for much longer: three to six months.

We also have our Baywell, which is a soft, creamy, but quite mature-tasting cow's milk cheese, made in a heart shape with an orangey-pink rind. It is an extension of the ripe, creamy-soft Penyston, named after Penyston Hastings, whose family owned Daylesford in the eighteenth century – and which we recently revived. As often happens in artisan cheesemaking, the Baywell came about almost by accident, when someone tried rind-washing a troublesome batch of Penyston. Washing the rind simply means that the cheese is washed in brine during its maturation to inhibit any mould growth and encourage the growth of certain bacteria instead. It's these bacteria that give rind-washed cheeses their pungent smell and distinctive flavours.

We have also developed blue cheeses. The Bledington Blue cheeses are small, creamy versions of our Daylesford Blue, in which the curds are stirred for less time so that they are softer and stay in salt for slightly less time, then the cheese is matured for just four to nine weeks.

Some people say that true Cheddar can only be made in Somerset, but it is the process that makes a Cheddar, although every dairy makes it slightly differently. Some make it slightly wetter than others, and obviously everyone is starting with different milk – in our case the milk comes from our own Friesian herd.

Once the coagulated milk is cut to separate the curds and whey, the whey is drained off and the mass of curds comes together in springy pillows which are layered up, left for a while, then turned – this is the 'cheddaring' process, which we repeat around five times to press and drain the curd and let the acidity rise, then it is shredded and salted. It is then packed into the moulds, which are lined with cotton muslin and pressed. It's an extraordinary process really, achieving something solid out of soft curds. People think that the pressing is to get the moisture out but not at all; that has already happened and the pressing is to shape the cheese and allow it to come together. Traditional clothbound Cheddar, being a hard cheese and one of the driest, needs more pressure than most. We still use the old wooden presses that work with ropes, pulleys and weights.

We aim to mature our Cheddar for nine months or more. I like it when it's been aged for longer – between ten and twelve months. We do age some of our cheeses for eighteen months, and they can be wonderful, with a more nutty flavour, but as soon as you cut into an older cheese, it will start to dry out, so you can't keep it for long. Only cheeses of a certain character are suitable for longer ageing. You check each one at around nine months using a stainless steel cheese iron, which is pushed into the centre and pulls out a plug of cheese, so you can assess the aroma, body, texture and flavour.

In many big dairies, they use a different culture and most of the cheese-making is done by machine, ending with the cheese being forced down a long tube and into its plastic wrapper. It can be very good cheese, but it tends to have a sweeter taste that isn't a true Cheddar flavour. People who have got used to this are sometimes surprised when they taste a traditional hand-made cheese.

At Daylesford, the creamery is next door to the bakery and we share the same philosophy: bread and cheese have always gone together, and they both need time and skill to make them well. You can churn out bread very fast in a factory, using the Chorleywood process, or you can make a sourdough slowly, giving it time to develop; just as you can produce tons of blocks of cheese in a day or you can make single batches and nurture them all the way through the process.

SAVOURY TARTS & PIES

Notes on pastry

The tarts made at Daylesford change continually as we move through the seasons. The chefs combine vegetables from the market garden with different cheeses from the creamery – packing the pastry cases quite full with whatever produce we are using, so that there is a high ratio of vegetables to cheese.

The chefs tell me that our tarts are baked at a very low temperature, so that the eggs don't expand and cause the filling to rise up, like a soufflé, as if this happens the mixture will drop again as the tart cools, and the surface will crack. Instead the more gentle baking allows the filling simply to set. Also, these are deep tarts, so if you use too high a temperature, the top will be brown before the mixture has cooked in the centre.

In autumn, especially when local game comes into the kitchen and we move into much more slow casserole-style cooking, it is time to start making homely, comforting pies in deep dishes, the kind in which the pastry is held up by an old-fashioned pie prop which pops through the centre of the golden top. These are perfectly partnered with winter brassicas or crushed, buttered root vegetables and cabbage (see page 127).

Savoury Pastry

This makes enough for two of the savoury tarts on the following pages (each one 20cm round and around 5–7cm deep) or one each of the Cheddar, potato and onion pie on page 169 and the Wootton Estate game pie on page 179. If you are only making one tart, the remaining pastry can be frozen and used up to two months later – and, of course, you can double the quantity and make a bigger batch, then portion it and put it into the freezer for whenever you need it.

370g plain flour	265g butter, cold
a pinch of salt	1 egg (plus 1 egg, beaten, for brushing)

Sift the flour into a bowl and add the salt. Grate in the butter and mix lightly with the tips of your fingers until the mixture resembles breadcrumbs, ensuring there are no lumps of butter in the mix (alternatively you can do this using a food processor).

Crack the egg into a measuring jug, beat lightly and top up with cold water until you have 140ml. Pour this slowly into the flour and butter, mixing for a few seconds until the mixture forms a dough. Don't overwork it or the pastry will be tough. Divide into 2 balls, wrap in clingfilm and chill in the fridge for at least 30 minutes before using. (Or freeze until needed).

Preheat the oven to 160°C/gas 3.

Lightly flour your work surface and then, for each tart, roll out one of your balls of pastry into a circle big enough to line a 20cm round by 5–7cm deep flan tin with a removable base, leaving enough pastry to overhang the sides. Wrap the pastry carefully around your rolling pin to lift it and drop it carefully into the flan tin, pushing it gently into the base and sides of the tin – don't trim the overhanging pastry. Put the tin on a baking tray – this makes it easier to move it around – then into the fridge to rest and chill for another 30 minutes (to help prevent the pastry shrinking during baking).

Savoury Pastry continued

When ready to blind-bake, prick the base of the pastry case with a fork, line with greaseproof paper – crinkle it up first to soften it and avoid it denting the pastry – and fill with baking beans. Put into the oven for about 30 minutes, until light golden brown, then take out, remove the paper and baking beans (you no longer need these), and brush all over the inside of the pastry case with the beaten egg, to seal any little holes.

Put the tin back into the oven for a further 5–10 minutes, until the base is fully baked and golden brown. Don't be scared of taking the pastry to this point. The key to a good tart base is to hold your nerve, and colour and crisp the pastry to the stage at which you would like to eat it, as once you put in your filling and return it to the oven it won't colour any more, except maybe a little around the edges and the base will stay crispy and flaky as the filling cooks. If you only lightly colour the pastry, and the base isn't fully baked, it will be soft and doughy, making the whole tart seem heavy.

Remove from the oven and, when cool, carefully trim off the overhanging pastry with a small, sharp, serrated knife. Keep to one side. Now you can make whichever filling you like and bake your tart according to the instructions in each recipe.

SERVES 6

Adlestrop and Kale Tart

Our Adlestrop cheese, which takes its name from a local village, is a washed-rind, quite pungent, semi-soft cheese, matured for 10 weeks. You could also use a cheese like Caerphilly.

For this tart, the Daylesford chefs usually use a heritage variety of kale called Red Russian, which is green with a reddish tinge around the edges and is a little flatter than the more common varieties. Cavolo nero works well, too.

- 80g butter
- 2 small white onions, finely chopped
- 2 cloves of garlic, finely chopped
- 250g kale leaves, stalk removed and leaves shredded
- 2 eggs
- 4 egg yolks
- 250g mascarpone
- 100ml double cream
- 250g Adlestrop or similar cheese, chopped (about 2cm)
- 1 blind-baked 20cm x 5–7cm shortcrust tart case (see page 153)
- sea salt and freshly ground black pepper

Preheat the oven to 140°C/gas 1.

Melt the butter in a large pan, then put in the onions and garlic and cook over a low heat for 5 minutes, until the onions have softened but not coloured. Add the kale and cook for another 5 minutes, stirring constantly, until it has just slightly softened. Taste and season as necessary. Transfer to a bowl and leave to cool.

In a separate, large bowl, whisk the eggs, yolks, mascarpone and cream for a few seconds, then add the cheese. Squeeze the excess water from the kale and onions, and add to the mixture. Stir in lightly and spoon into the prepared pastry case, making sure the cheese is evenly distributed. Put into the oven and bake for about 40 minutes, or until the top is golden and the mixture is fully cooked – to check, insert a metal skewer into the centre and if it doesn't smear, the tart is done.

Goat's Cheese and Asparagus Tart

English asparagus is paired with a goat's cheese that is semi-soft and creamy, not too fresh, and not too strong: Windrush, Ragstone, Golden Cross and Crottin de Chèvre are all good choices.

250g asparagus

2 eggs

4 egg yolks

250g mascarpone

100ml double cream

1 small bunch of chervil, chopped

1 blind-baked 20cm x 5–7cm shortcrust tart case (see page 153)

230g fresh goat's cheese, such as Oddington

sea salt and freshly ground black pepper

Preheat the oven to 140°C/gas 1.

Unless your asparagus is young and tender, take off the white woody bases and slice the spears very thinly crossways, leaving just the very tips intact.

In a large bowl, whisk the eggs, yolks, mascarpone and cream for a few seconds, then add the asparagus, chervil and seasoning, mix lightly and spoon into the prepared pastry case.

Crumble the goat's cheese and dot over the surface, so that some pieces stick out. Put into the oven and bake for about 40 minutes, or until the top of the tart is golden and the mixture is fully cooked – to check, insert a metal skewer into the centre and if it doesn't smear, the tart is done.

SERVES 6

Bledington Blue Cheese and Broccoli

Our Bledington Blue is a softer-flavoured, less crumbly but slightly more creamy cheese than, say, a Stilton. You could use Barkham Blue, Cambozola or Stichelton, depending on what you can get hold of.

- 2 heads of broccoli, separated into florets
- 2 eggs
- 4 egg yolks
- 250g mascarpone
- 100ml double cream
- 3 tablespoons chopped fresh chervil
- 1 blind-baked 20cm x 5–7cm shortcrust tart case (see page 153)
- 150g blue cheese, such as Bledington
- sea salt and freshly ground black pepper

Preheat the oven to 140°C/gas 1.

Two-thirds fill a large pan with cold water and bring to a rapid boil. Add the broccoli and cook over a high heat for 2 minutes, then take off the heat and drain through a colander under cold running water until cold. Leave the broccoli to stand in the colander for 10 minutes, then pat dry with a clean tea towel or kitchen paper.

In a large bowl, whisk the eggs, yolks, mascarpone and cream for a few seconds, then add the cooked broccoli and chervil, season, mix lightly and spoon into the prepared pastry case.

Crumble the blue cheese and dot over the surface, so that some pieces stick out. Put into the oven and bake for about 40 minutes, or until the top is golden and the mixture is fully cooked – to check, insert a metal skewer into the centre and if it doesn't smear, the tart is done.

SERVES 4

Red Onion Tarte Tatins with Baywell Cheese

Baywell is a rind-washed cheese that is soft and creamy but quite mature in flavour. You could try Camembert, Brie, Adlestrop or Crottin goat's cheese in its place. The caramelised onions serve as a savoury alternative to the traditional apples in these small tarts, which we make in 12cm blini pans that can be transferred to the oven. You could use similar-sized individual tart tins.

6 small red onions (try to choose onions all the same size), peeled and left whole

1 tablespoon olive oil

a couple of knobs of butter

1 tablespoon balsamic vinegar

20g sugar

3 tablespoons red wine

1 x 200g pack of puff pastry

1 Baywell cheese, cut into 8 slices, leaving the rind on

4 handfuls of mixed salad leaves, to serve

For the dressing:

3 tablespoons olive oil

2 tablespoons balsamic vinegar

sea salt and freshly ground black pepper

Slice each onion exactly in half through the root, to give two identical, intact halves from each onion – so you end up with 12 halves of the same size.

Preheat the oven to 180°C/gas 4.

Heat the oil and butter in a large, non-stick and ovenproof frying pan. When the butter is foaming, put in all the onions, cut side down, and cook gently until the cut sides are golden brown. Mix together the vinegar, sugar and red wine, then add to the pan and bring to the boil. Cover the pan with foil, transfer to the oven and bake for 15 minutes, or until the onions are tender.

Meanwhile, lightly flour a work surface and roll out the pastry to about 3–4mm thickness. Cut 4 rounds slightly larger than your blini pans or tart tins and layer on a plate in the fridge until needed.

Remove the pan of onions from the oven and carefully lift out the onion halves. Divide the cooking liquor between each pan or tin, and then arrange 3 onion halves, cut side down on top.

Drape a chilled pastry top over each trio of onion halves, tucking in the pastry around the insides of the pan or tin. Put on a baking tray and bake in the oven for about 20–25 minutes, or until the pastry is well risen and golden brown.

Meanwhile, mix the olive oil and balsamic vinegar together for the dressing, season and toss through the mixed leaves.

Remove the tarts from the oven and turn out on to four plates – to do this, hold each plate firmly over the top of a pan or tin and flip both over together, so that the tart ends up on the plate with the onion side upwards. Top each one with 2 slices of cheese (this will melt) and serve with some of the dressed leaves.

Butternut Squash and Kale Tart

SERVES 6

Another creamy tart with earthy autumnal flavours. You need to cook and drain the greens well and remove any moisture before adding them to the filling otherwise it will be thin and might seem watery.

- 80g butter
- 100g kale leaves, shredded
- 1 medium butternut squash, peeled, halved, seeds removed and flesh grated
- 2 eggs
- 4 egg yolks
- 250g mascarpone
- 100ml double cream
- 70g Cheddar cheese, grated
- 100g Parmesan cheese, finely grated
- 1 blind-baked 20cm x 5–7cm shortcrust tart case (see page 153)
- sea salt and freshly ground black pepper

Preheat the oven to 140°C/ gas 1.

Melt the butter in a medium pan, add the kale and cook for 2 minutes, stirring constantly. Increase the heat, add the grated butternut squash and continue to cook for 2–3 minutes, until the vegetables are just slightly softened. Taste and season accordingly, then take the pan from the heat and turn the mixture into a bowl or onto a plate to cool.

In a separate bowl, whisk the eggs, yolks, mascarpone and cream for a few seconds, then add the Cheddar and Parmesan. Drain any excess moisture from the kale and squash mixture, season again to taste, and mix lightly with the egg mixture. Spoon into the prepared pastry case and smooth level with the back of the spoon.

Put into the oven and bake for about 40 minutes, or until the top of the tart is golden and the mixture is fully cooked – to check, insert a metal skewer into the centre and if it doesn't smear, the tart is done.

SERVES 6

Jerusalem Artichoke and Cavolo Nero Tart

Jerusalem artichokes are always sweeter and more flavoursome after the first frost so this is a tart for the winter months.

- 700g Jerusalem artichokes
- 3 tablespoons olive oil
- 2 tablespoons butter
- 300g bunch of cavolo nero, stalks removed and leaves finely chopped
- 2 eggs
- 4 egg yolks
- 250g mascarpone
- 100ml double cream
- 100g Parmesan cheese, finely grated
- 1 x blind-baked 20cm x 5–7cm shortcrust tart case (see page 153)
- sea salt and freshly ground black pepper

Preheat the oven to 180°C/gas 4.

Cut the top and bottoms off the artichokes and wash them well in a bowl of water, using a nail scrubbing brush to remove all the sand and dirt. Cut into quarters, put into a small bowl with the olive oil and season with a little salt and pepper. Toss well to coat, then transfer to a baking tray and put into the oven for about 12 minutes, until softened and golden brown. Remove from the oven and leave to cool. Turn the oven down to 140°C/gas 1.

Melt the butter in a medium pan, then put in the cavolo nero and cook for 5 minutes over a medium heat. Season and leave to cool, then drain off any buttery liquid, transfer to a board and chop a little more finely.

In a large bowl, whisk the eggs, yolks, mascarpone and cream for a few seconds, then add the Parmesan, chopped cavolo nero and roasted artichokes. Season and mix lightly, then spoon into the prepared pastry case and smooth level with the back of the spoon. Put into the oven and bake for about 40 minutes, or until the top is golden and the mixture is fully cooked – to check, insert a metal skewer into the centre and if it doesn't smear, the tart is done.

MAKES 6

Three Tomato Tart

This tart rides and falls on the quality of your tomatoes. Jez recommends Marmande, which are the juicy, sweet, 'ribbed', ones that have more flesh than seeds – they are good for the sauce and for slicing. Earlier in the year Stupice is one of the first varieties to ripen in our market garden and is a good, full-flavoured one to use for the sauce. Sweet, juicy Golden Cherry tomatoes are perfect for the finishing layer.

12 large ripe red vine tomatoes, preferably heritage (see introduction, above)

6 tablespoons extra virgin olive oil

1 medium red onion, thinly sliced

2 cloves of garlic, thinly sliced

1 teaspoon coriander seeds

1 teaspoon cumin seeds

1 teaspoon tomato paste

1 teaspoon chopped fresh red chilli

juice of ½ a lemon

1 good sprig of thyme, leaves only

500g puff pastry

3 tablespoons freshly grated Parmesan cheese

20 yellow vine cherry tomatoes, preferably heritage (see introduction, above)

sea salt and freshly ground black pepper

Preheat the oven to 180°C/gas 4.

Chop 6 of the large tomatoes.

Heat 3 tablespoons of the olive oil in a medium pan over a low heat, then add the onion, garlic and the coriander and cumin seeds, and cook for about 10 minutes, until the onion softens, stirring constantly. Add the chopped tomatoes, tomato paste, chilli, lemon juice and half the thyme leaves and continue to cook for 5 minutes.

Add 200ml of water and bring to the boil, then cook over a medium heat for about 15 minutes, until the tomatoes have broken down and all the juice has evaporated. Taste and season as necessary. Remove from the heat and leave to cool.

Roll out the puff pastry on a lightly floured surface to about 3–4mm thick and cut around a small plate or use a large cutter (about 12cm in diameter) to cut out 6 discs. Line a baking tray with a sheet of greaseproof paper, then place the discs on top and put into the fridge

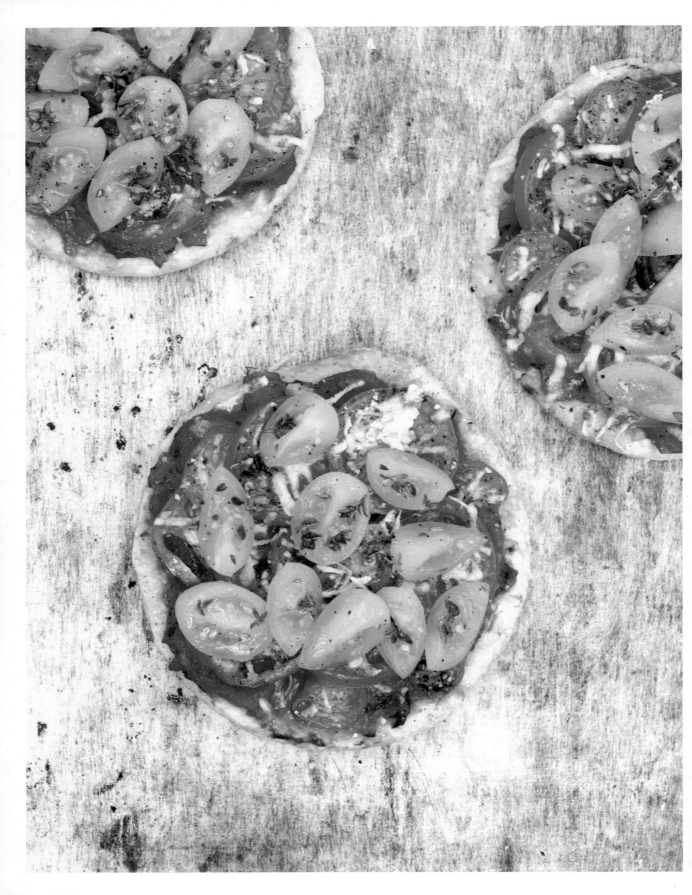

Three Tomato Tart continued

to rest for 20 minutes (this will help to stop the pastry shrinking in the oven).

When ready to cook, remove the pastry discs from the fridge and prick all over with a fork. Bake in the oven for 15 minutes, until light golden and crisp.

Remove and leave to cool. Slice the remaining red tomatoes thinly (around 3mm).

Divide the cooked tomato mixture between the pastry discs, spreading it right to the edge. Arrange the slices of red tomato on top (approximately 6 slices on each). Sprinkle with Parmesan and return the tarts to the oven for a further 10 minutes.

Meanwhile, cut the yellow cherry tomatoes into halves or quarters depending on their size, put into a bowl, combine with the rest of the olive oil and the remaining thyme leaves and season.

Remove the tarts from the oven, spoon some of the yellow tomatoes and their juices on to each tart and serve.

SERVES 6

Single Gloucester, Spinach and Smoked Bacon Tart

Single Gloucester's gentle flavour works with the strong smoky bacon but we are one of only a handful of creameries who make it so if you can't get hold of it, other cheeses that combine well with the spinach and bacon include Adlestrop and Caerphilly.

- 50g butter
- 150g smoked streaky bacon, cut into small pieces
- 1 onion, finely chopped
- 450g baby spinach
- 2 eggs
- 4 egg yolks
- 250g mascarpone
- 100ml double cream
- 150g Single Gloucester cheese, grated
- 1 blind-baked 20cm x 5–7cm shortcrust tart case (see page 153)
- sea salt and freshly ground black pepper

Preheat the oven to 180°C/gas 4.

Melt the butter in a large non-stick frying pan over a medium heat. Put in the bacon pieces and cook for about 5 minutes, until golden brown (but not crispy), stirring constantly, then lower the heat, add the onion and continue to cook for about 10 minutes, until the onion has softened but not coloured.

Add the spinach and continue to cook for 1 minute or until it just wilts, then take off the heat immediately, drain through a sieve and transfer to a bowl to cool.

In a separate, large bowl, whisk the eggs, yolks, mascarpone and cream for a few seconds, then add the cheese. Squeeze any remaining moisture from the spinach mixture and add to the bowl, season and mix lightly, then spoon into the prepared pastry case and smooth level with the back of the spoon.

Put into the oven and bake for about 35 minutes, or until the top is golden and the mixture is fully cooked – to check, insert a metal skewer into the centre and if it doesn't smear, the tart is done.

SERVES 8

Cheddar, Potato and Onion Pie

This needs nothing more than a simple side of steamed greens or a mixed leaf salad to accompany it. The green beans with almonds, parsley and garlic butter on page 124 work well.

- 1 quantity of savoury pastry (see page 153)
- a little plain flour, for rolling out
- 500g (peeled weight) potatoes, cut into cubes (about 2cm)
- 2 medium onions, chopped
- 600g Cheddar cheese, coarsely grated
- 3 medium eggs, beaten
- 1 medium egg yolk, beaten

To make this, you need a flan tin with a removable base, 25cm in diameter and around 5–7cm deep.

Lightly flour your work surface. Take around two-thirds of the pastry (wrap the rest in clingfilm and chill in the fridge) and roll out into a circle about 3mm thick and 5cm larger than the tin. Wrap the pastry carefully around your rolling pin to lift it and drop it gently into the flan tin. Press the pastry into the base and sides of the tin, leaving a little excess overhanging the edge.

Put the tin on a baking tray and chill in the fridge while you prepare the filling.

Half fill a medium pan with lightly salted water and bring to the boil. Add the potatoes and return to the boil. Cook for 5 minutes, or until cooked but still slightly firm. Add the onions and return to the boil, then drain immediately in a colander. Tip into a large bowl and leave to cool.

When the vegetables are cold, stir in the cheese and beaten whole eggs until thoroughly combined.

Take the pastry case from the fridge and fill with the potato and cheese mixture.

Cheddar, Potato and Onion Pie continued

Lightly flour your work surface again and roll out the remaining pastry into a circle around 3mm thick and big enough to cover the pastry case, leaving a little overhang. Brush the edges of the pastry case with the beaten egg yolk, then wrap the pastry circle over your rolling pin and carefully drape it over the top. Press to seal the edges, then trim off the excess pastry. Make a large hole in the centre of the pie and brush all over the pastry lid with the remaining egg yolk. Put back into the fridge on a baking tray for an hour to chill and set – this will help to prevent the pastry from shrinking in the oven.

Preheat the oven to 180°C/gas 4. To decorate the top, take a sharp knife, and with the tip gently score the pastry in gently curved lines from the centre to the sides of the pie, to resemble the spokes of a wheel.

Bake on the baking tray in the oven for 45–60 minutes, until the pastry is nicely browned and the filling is hot throughout – you can check by inserting a skewer into the centre of the pie. It should be piping hot when you remove it.

SERVES 6

Chicken, Leek and Bacon Pie

This recipe dates back to the very first days of the café in our farmshop in Kingham and is based on the Scottish cock-a-leekie soup combination of chicken and leeks. Ours includes a little smoked bacon to give the dish a richer depth of flavour. This is best made with small, young, tender leeks.

- 2 potatoes, peeled and finely chopped
- 2 tablespoons sunflower oil
- 3 onions, finely chopped
- 2 sticks of celery, finely chopped
- 1 teaspoon chopped fresh thyme leaves
- 50g smoked bacon lardons
- 1 teaspoon sugar
- 800g raw chicken, cut into bite-size pieces (use a mixture of thigh/leg and breast meat)
- 50g plain flour, plus extra for rolling out
- 1 tablespoon lemon juice
- 800ml good chicken stock
- 2 leeks, sliced
- 3 tablespoons crème fraîche
- 1 tablespoon chopped fresh flat-leaf parsley leaves
- ½ quantity of savoury pastry (see page 153)
- 1 beaten egg, for brushing the pastry
- sea salt and freshly ground black pepper

Preheat the oven to 200°C/gas 6.

Half fill a medium pan with lightly salted water and bring to the boil. Add the potatoes and return to the boil. Cook for about 10 minutes, or until tender but still holding their shape. Drain and set aside.

Heat the oil in a pan, put in the onions, celery, thyme and bacon, together with a pinch of salt and the sugar, and cook gently until the onions are soft but not coloured.

Add the chicken and again cook gently without colouring for about 10 minutes. Add the flour, then slowly stir in the lemon juice and chicken stock. Bring slowly to the boil, then turn down the heat and simmer until the sauce has thickened.

Add the leeks and continue cooking until they are tender, then gently stir in the cooked potatoes. Take off the heat and stir in the crème fraîche and parsley.

Taste and season as necessary, then spoon into a round pie dish, roughly 20cm in diameter (or equivalent) and leave to cool. Put a pie prop in the middle.

Lightly flour your work surface and roll out the pastry to the shape of your pie dish, allowing enough for a good overhang all round.

Wrap the pastry over your rolling pin and carefully drape it over the top of the dish. Make a large hole in the centre of the pie above the prop and let it push through, then brush all over with the beaten egg.

Put into the oven for 30–35 minutes, until the pastry is golden brown and the meat is piping hot.

MAKES 6 INDIVIDUAL PIES

Venison and Cranberry Pies

These are quite rich pies but the tartness of the cranberries really lifts the flavour and helps cut through the richness.

You can quickly and easily cut out six tops from two sheets of good, ready-rolled puff pastry. However, if you prefer to make one big pie (approximately 25cm round or 20cm x 30cm if you have an oblong-shaped dish) it is easier to buy a block of pastry and roll it out yourself on a floured surface to the size needed. A large pie will need a pie prop placed in the middle of the venison mixture before you drape your pastry over the top (make a hole in the centre for it to poke through) and about 45 minutes in the oven.

The pies can be made up to a few hours in advance of cooking – just keep them in the fridge until you're ready to bake. If you do this, however, bring them up to room temperature before cooking.

1kg venison haunch or shoulder, cut into bite-size pieces

375ml red wine

2 medium carrots, chopped

2 medium onions, sliced

1 bay leaf

2 juniper berries

1 sprig of fresh thyme

50g butter, plus a little extra if needed

2 tablespoons olive oil

25g sugar

40g plain flour

1 tablespoon tomato purée

500ml good chicken stock

100g sun-dried cranberries

2 sheets of ready-made puff pastry (see introduction, above)

1 egg, beaten

sea salt and freshly ground pepper

Put the venison into a bowl with the red wine, carrots, onions, bay leaf, juniper berries and thyme and leave to marinate in the fridge for 24 hours.

When ready to make the pies, preheat the oven to 200°C/gas 6.

Remove the venison from the marinade and pat dry with kitchen paper. Strain the liquid into a bowl and keep to one side. Discard the juniper berries and thyme sprig, but keep back the carrots and onions.

Heat the butter and oil in a casserole. Put in the venison and sauté until golden brown all over, then lift out and reduce the heat. Add the reserved sliced onions and carrots from the marinade, together with the sugar, and soften the vegetables gently, adding a little more butter if needed.

Add the flour and tomato purée and cook for 1–2 minutes more, then add the marinade liquid a little at a time, stirring continually to prevent lumps from forming.

Return the browned meat to the pan, together with the stock, bring to the boil, then turn down the heat and simmer for about 2 hours, covered, until the venison is cooked and tender, and the sauce has thickened.

Stir in the cranberries, taste and season as necessary, then divide between six individual pie dishes.

Cut out 3 lids from each sheet of pastry – they should be big enough to overhang the dishes by at least 1cm. Drape over the top and brush with the beaten egg.

Put the dishes on a baking tray in the oven for 30 minutes, or until the pastry is risen and golden brown.

MAKES 1 DEEP PIE

Wootton Estate Game Pie

Wootton is our family farm up in Staffordshire. The landscape up there is very different to Daylesford – it's wild and rugged and less suited to arable farming but we have a wonderful herd of deer. It's the largest herd of organic farmed deer in the country, so venison is something we'll have on the butchers' counters at the farmshops throughout the winter. Our venison is very lean so it works well in a pie, which helps to conserve the moisture in the meat.

For the duck, we use mallard, which comes from a nearby estate, as this is a smaller duck with a gamier flavour than the milder-tasting Gressingham or Barbary breeds.

You need to allow time to make this, as the game meat needs to be marinated overnight, then cooked slowly for 4 hours, and allowed to cool and chill in the fridge for another 2 hours before you assemble the pie. This chilling of the meat might seem excessive, but it is important because you will have lined your pie dish and rolled out your lid ready to go, then put these into the fridge to chill and relax and so avoid any shrinking in the oven. So if you were to put warm meat into the chilled pie casing it would melt the pastry.

2 pheasant (or partridge) legs
2 duck legs
2 rabbit legs
500ml red wine
2 juniper berries
5 peppercorns
½ a head of garlic
5 sprigs of fresh thyme
250g diced haunch of venison
2 tablespoons olive oil
1 onion, chopped

2 sticks of celery, chopped
2 carrots, chopped
2 litres good chicken stock
1 celeriac, chopped
75g dried cranberries
1 quantity of savoury pastry (see page 153)
flour, for rolling out
1 egg, beaten, for brushing the pastry
sea salt and freshly ground black pepper

Put the pheasant, duck and rabbit legs into a large bowl with the wine, juniper berries, peppercorns, garlic and one of the sprigs of thyme. Cover and put into the fridge to marinate overnight.

Next day, preheat the oven to 150°C/gas 2.

Lift out the legs from the marinade and pat the meat dry with kitchen paper. Strain the liquid into a bowl and keep to one side. Discard the juniper berries, peppercorns, garlic and thyme.

Season the legs and also the venison. Heat the olive oil in a casserole (one that has a lid), put in the legs and the venison and sauté until browned on all sides. Lift out and keep to one side. Add the onion, celery and carrots and gently soften until browned, stirring occasionally.

Add the marinade and bubble up, scraping all the bits from the bottom of the casserole, until the alcohol has burned off. Return the legs and the venison to the casserole and cover with the chicken stock. Add the rest of the thyme.

Put the lid on the casserole and transfer to the oven for about 4 hours, until the meat is completely tender and the leg meat is falling away from the bones.

When the meat is tender, remove the casserole from the oven. Lift out all the meat and put it into a bowl to chill slightly (leave the cooking liquor in the casserole). When cool enough to handle, strip the meat from the bones (discard the bones). Keep the meat to one side.

Put the casserole dish containing the cooking liquor on the hob over a medium heat and bubble up to reduce by half. Add the celeriac and cranberries and simmer for 10 minutes, until the cranberries have softened and the celeriac is tender. Then put back the reserved meat. Stir gently, remove from the heat and leave to cool down, then put into the fridge to chill for 2 hours.

While the meat is chilling, line your pie dish. You need a dish of around 20cm in diameter (or equivalent). Lightly flour your work surface. Take around two-thirds of the pastry (wrap the rest in clingfilm and chill in the fridge), and roll out into the shape of your pie dish, but around 5cm larger. It needs to be about 4mm thick. Wrap the pastry carefully around your rolling pin to lift it, then drop it gently into the dish. Press into the base and sides, leaving a little excess overhanging the edge. Put the dish on a baking tray and chill in the fridge until ready to bake.

For the lid of the pie, lightly flour your work surface again and roll out the remaining pastry into the shape of your dish, but big enough to leave a little overhang – again it should be around 4mm thick. Lay this on a plate or tray and put into the fridge to chill.

When ready to bake, preheat the oven again to 180°C/gas 4. Take the pastry-lined dish from the fridge and spoon in the chilled meat mixture. Place a pie prop in the middle.

Brush the edges of the pastry in the dish with the beaten egg yolk, then wrap the chilled pastry for the lid over your rolling pin and carefully drape it over the top. Press to seal the edges, then trim off the excess pastry. Make a hole in the centre of the pie above the prop and let it push through, then brush all over the pastry lid with the remaining egg yolk.

Put the pie into the oven and bake for 1 hour, or until the pastry is golden brown and the filling is piping hot.

Notes on fish

Visitors to our farmshops and kitchens like to buy and eat fish that they can trust to be sustainably sourced. But of course we are in the middle of the Gloucestershire countryside, and farming livestock and growing fruit and vegetables is what we do – we don't fish ourselves. So we bring in our fish from trusted suppliers around the coast, mainly Cornwall, who know to supply us only with a catch that meets the stringent guidelines drawn up by Tim Field, our environmental scientist, who is also a dedicated wild food hunter and forager. In fact, says Tim, in the River Evenlode, which runs alongside the wetland area he has created for the farm, 'there is a lovely little population of wild brown trout and I am sure they would taste beautiful, but they are all about 12cm long – not even big enough for breakfast.'

There are crayfish in the Evenlode too, but as in most British rivers they are the rampant American red claw or 'Signal' species, which are quite territorial and aggressive and have largely driven out the population of white-clawed British crayfish. They also carry a disease that doesn't affect them, but wipes out the natives. 'Catching crayfish is a wonderful experience', says Tim. 'I was brought up locally and my brother and I used to catch them virtually with a jam jar, bare hands and lightning reflexes, but these days you can only fish crayfish under licence. It's to keep tabs on the populations and also to try and prevent the intruders and their disease spreading. So lovely as it is to catch the odd crayfish, it would be terrible to think we were contributing to the further demise of the white-clawed ones.'

'Trying to fish sustainably is a very complex, constantly changing subject,' says Tim, 'as the sea is a dynamic environment and fish move around a great deal from warm to cold seas and rivers at different times of the year. You have to look at the seasons in terms of breeding and migrating, as well as the method of fishing, to avoid damage to the seabed habitat, sea birds, juvenile fish and non-targeted species caught as by-catch.

'Our wild fish comes primarily from small dayboats and members of the Responsible Fishing Scheme, so we know the stocks are healthy and we can be confident we are not damaging either the fish populations themselves or the ocean environment. The simple way to choose wisely is to check the Marine Conservation Society's Good Fish Guide (www.goodfishguide.co.uk) and Fishonline (www.fishonline.org).' Because of the issues involved, we put relatively few fish dishes on the menu. These, however, are some of the favourites.

SERVES 4

Grilled Mackerel with Roasted Beetroot and Spiced Lentils

The pairing of oily mackerel with earthy, sweet beetroot has become something of a British classic, but this recipe also reflects the influence of chefs of different nationalities who have worked in the Daylesford kitchens over the years, with the hint of chilli, ginger, sesame and soy sauce in the creamy lentils, and the fresh citrus, mint and coriander flavours in the yoghurt dressing.

Because omega-3-rich mackerel is so good for you and we have all been urged to eat more of it, there are some fisheries which have been put on the Fish to Avoid list; however, all our mackerel is sustainably fished from dayboats using hand lines off the Cornish or Devon coast. Ask your fishmonger to pin-bone the fish for you.

The pomegranate vinegar used to dress the beetroots is a bit special – you can really taste the pomegranate – but aged red wine vinegar is fine – you just want a touch of acidity to balance the sweetness of the beetroot, so don't use balsamic vinegar.

Although the chefs generally use sea salt in their cooking, as it dissolves quickly, for the yoghurt dressing they use rock salt, as it is harder and crushes better with the spices.

8 fillets of mackerel, pin-boned

2 tablespoons olive oil, plus a little extra virgin olive oil to finish

sea salt and freshly ground black pepper

For the spiced lentils:

230g dried green lentils

1 large red chilli, finely chopped

3cm piece of fresh root ginger, peeled and finely chopped

3 tablespoons olive oil

3 tablespoons sesame oil

2 tablespoons soy sauce

2 tablespoons white wine vinegar

2 tablespoons chopped fresh coriander

For the roasted beetroot:

500g beetroot, unpeeled but washed

2 tablespoons olive oil

2 tablespoons pomegranate vinegar or aged red wine vinegar

For the dressing:

1 teaspoon coriander seeds

1 teaspoon cumin seeds

½ teaspoon rock salt

150g natural yoghurt

50g crème fraîche

zest of ½ a lemon

zest of ½ an orange

1 tablespoon chopped mint

1 tablespoon chopped coriander

Grilled Mackerel continued

First prepare the lentils. Rinse them carefully, then cover with 750ml of cold water. Add the chilli and ginger, bring to the boil, then turn down the heat and simmer until the water is absorbed and the lentils are tender, but still retain a little bite. Leave to cool, then add the olive oil, sesame oil, soy sauce, vinegar and coriander. Season well and keep to one side.

While the lentils are cooking, put the beetroot into a medium-sized pan and cover with cold water. Bring to the boil, then turn down the heat and simmer for about 45 minutes, until tender. Drain, peel and roughly chop into bite-size chunks. Mix with 1 tablespoon of the olive oil, season and spread on a baking tray.

Preheat the oven to 170°C/gas 3. Put the beetroot into the oven for 10 minutes, so that the pieces roast a little and their flavour will be concentrated, then tip into a small bowl and add the vinegar and the rest of the oil. Toss together lightly.

To make the dressing, lightly toast the seeds in a small dry pan, just enough to release their aroma, then transfer them to a pestle and mortar, add the rock salt and crush to a fine powder. In a bowl, mix the yoghurt, crème fraîche, lemon and orange zest and herbs, then add the crushed seeds and salt and stir. Taste and add a little more salt if necessary.

Preheat the grill to hot. Brush the mackerel with the olive oil, season, then place under the grill, skin side up, and cook for about 5 minutes, until the skin is golden brown and the flesh is cooked through (it will turn opaque) but is still moist. Serve with the lentils and roasted beetroot and drizzle with some extra virgin olive oil and a good dollop of yoghurt dressing.

SERVES 4

Pan-roasted Pollock with Crushed Potatoes and Watercress Mayonnaise

Cooking with pollock is easier on fish stocks than buying cod (unless this comes from a sustainable fishery), and it is very similar in texture and appearance, though it has a slightly less pronounced taste – however, the zingy flavours in the crushed potatoes and the pepperiness of the watercress mayonnaise combine to make this a really vibrant dish.

2 tablespoons rapeseed oil

4 x 150g pieces of pollock, skin on and pin-boned

1 tablespoon butter

juice of ½ a lemon, plus 4 lemon wedges

sea salt and freshly ground black pepper

For the crushed potatoes:

400g salad potatoes, peeled and cut into bite-size pieces

2 tablespoons olive oil

125g podded fresh peas (or frozen peas, defrosted)

4 spring onions, finely sliced

2 tablespoons finely shredded fresh basil

2 tablespoons finely shredded fresh mint

zest of 1 lemon

For the watercress mayonnaise:

1 bunch of watercress, large stalks removed

100g mayonnaise

zest of 1 lemon and the juice of ½ a lemon

Preheat the oven to 200°C/gas 6.

For the crushed potatoes, put the potatoes into a pan of cold, lightly salted water and bring to the boil, then turn down the heat and simmer for 10 minutes, until tender. Drain in a colander and return them to the pan, crush slightly with a back of a spoon, then mix in the olive oil, peas, spring onions, herbs and lemon zest. Season to taste.

While the potatoes are cooking, make the watercress mayonnaise. Have ready a bowl of iced water. Bring a pan of water to the boil, then dip in the watercress to blanch it for 5 seconds only, lift out with a slotted spoon or sieve, and put into the iced water to cool. Squeeze out the water, then chop and mix with the mayonnaise, lemon zest and juice. Season to taste.

Pan-roasted Pollock continued

To cook the pollock, heat the oil in a non-stick frying pan that will transfer to the oven. Season the fish and put into the pan, skin side down. Cook over a medium heat until the skin is crisp and golden, then turn over and put into the oven for about 6 minutes, until just cooked (it should be opaque all the way through).

Remove the pan from the oven, add the butter and lemon juice, let the butter melt, then spoon over the fish. Divide the fish, with their buttery juices, between four plates, and add a lemon wedge to each, together with some of the crushed potatoes and a good dollop of watercress mayonnaise.

SERVES 4

Salmon and Smoked Haddock Fishcakes

Combining smoked haddock and salmon gives these fishcakes an interesting dimension as smoked fish has a different texture to fresh (the same principle applies to fish pie, see page 192).

Our chefs don't believe in cooking the fish before mixing it with the mash because the cakes still have to be fried and then finished in the oven. If you start with it raw, the cakes will be crunchy on the outside, but the fish will still be moist inside.

You can buy a good tartare sauce to go with the fishcakes – or serve them on their own – but it is quick and easy to make.

525g floury potatoes, peeled and cut into chunks

350g salmon, skinned, boned and chopped (about 1cm)

100g smoked haddock, skinned, boned and chopped (about 1cm)

2 tablespoons capers, chopped

2 tablespoons chopped fresh flat-leaf parsley

zest and juice of ½ a lemon

4 tablespoons plain flour

1–2 medium eggs, beaten

about 100g dried white breadcrumbs

sunflower oil, for frying

sea salt and freshly ground black pepper

rocket leaves (optional), for garnish, dressed with French dressing (see page 335)

lemon wedges, to serve

For the tartare sauce:

1 tablespoon capers

1 tablespoon gherkins, finely chopped

150g mayonnaise

1 tablespoon chopped fresh flat-leaf parsley

Put the potatoes into a pan of cold, lightly salted water and bring to the boil, then turn down the heat and simmer until the potatoes are tender if pierced with the tip of a sharp knife, but not falling apart. Drain through a colander, put into a bowl, allow to cool, then lightly mash.

Add the chopped fish, capers, parsley, lemon zest and juice, season, and mix well. Divide the mixture evenly into 8 balls, then flatten them and shape them into round cakes on a board. Put them into the fridge (on a board or large plate) to chill for at least 1 hour to firm up.

Meanwhile, if you are making the tartare sauce, combine the capers, gherkins, mayonnaise and parsley, and season with pepper. Put into the fridge until you are ready to fry the fishcakes.

Preheat the oven to 180°C/gas 4.

Have ready the flour, beaten egg and breadcrumbs in separate, shallow bowls and season each one with salt and pepper. Dip the chilled fishcakes first into the flour, shaking off the excess, then into the egg and then finally into the breadcrumbs, pressing them in lightly all over.

Heat around 1–2mm of the sunflower oil in a frying pan and carefully lift in four of the fishcakes. Gently fry over a medium heat, cooking in two batches, until golden brown on both sides. As they cook, transfer them to a baking tray and, once they are all ready, place them in the oven for 10–12 minutes, until hot in the centre.

Arrange 2 fishcakes per person on each plate, with a wedge of lemon. Garnish, if you like, with rocket, and serve with the tartare sauce.

SERVES 6

Traditional Fish Pie

I love the soothing simplicity of a well-made fish pie – it's not fancy but it always pleases a crowd and is a good thing to serve at a relaxed supper with friends. Daylesford's chefs say that the key to a good pie is to keep the fish moist in its sauce and not to swamp it with too much potato, otherwise the pie will feel stodgy. You don't want to cook the fish too much or let the sauce become thick before you assemble the pie – the temptation is to think of it as being the way you want to eat it, but remember that the pie still has to go into the oven for about half an hour and the fish and sauce will dry out some more.

The capers add a bit of texture and a vinegary sharpness and saltiness that works well with the fish: think of fish and chips.

200g smoked haddock, skinned and boned

250g salmon, skinned and boned

250g pollock, skinned and boned

200g leeks, sliced

3 eggs

1.5kg potatoes, peeled and cut into 4 pieces

150g butter

200ml white wine

1 bay leaf

50g plain flour

100ml double cream

1 tablespoon Dijon mustard

1 tablespoon capers

1 tablespoon chopped fresh flat-leaf parsley

sea salt and freshly ground black pepper

Preheat the oven to 180°C/gas 4.

Cut all the fish into bite-size pieces.

Blanch the leeks in boiling water for 30 seconds. Remove to a colander with a slotted spoon, then drain under the cold tap and keep to one side.

Put the eggs into a pan and cover with cold water. Bring to the boil and cook for 10 minutes. Take off the heat, rinse the eggs immediately in cold water to prevent discolouring, then peel and grate or roughly chop.

Put the potatoes into a separate pan of cold, lightly salted water, bring to the boil, then turn down the heat and simmer until the potatoes are tender if pierced with the tip of a sharp knife, but not falling apart. Drain in a colander, then put into a bowl and mash well (alternatively put them through a potato ricer). Stir in 50g of the butter and season with salt and pepper.

Pour the wine into a pan, add 500ml of water and bring to the boil. Add the chopped fish and the bay leaf. Simmer for about 2 minutes, until the fish is cooked (it will turn opaque), then take the pan off the heat, lift out the fish and leave to cool, reserving the poaching liquid.

Melt the rest of the butter in a separate pan and whisk in the flour until smooth. Gradually add the reserved poaching liquid from the fish, then the cream, stirring constantly until thickened – but not too thick (see introduction, above). Season and add the mustard. Gently stir in the cooked leeks, capers, parsley and reserved cooked fish.

Spoon the mixture into an ovenproof dish, top with the grated egg, spoon over the mashed potato (or pipe it, if you prefer) and put into the oven for about 30 minutes, until golden brown and piping hot.

Cod with Lemon, Parsley and Tomato Butter

SERVES 4

Much has been written about the depletion of cod stocks, but we buy our cod only from fisheries that are well managed and sustainable. Choose nice, fat pieces of fillet that have been cut from a big fish.

- 4 large red vine tomatoes or 20 cherry tomatoes
- 4 x 160g pieces of thick cod fillet (scales and pin-bones removed)
- 3 tablespoons olive oil
- 2 lemons and the juice of 2 more lemons
- 6 tablespoons butter
- 4 tablespoons capers
- ½ a bunch of fresh flat-leaf parsley, leaves roughly chopped
- sea salt and freshly ground black pepper
- 4 bunches of watercress, to serve (optional)

Preheat the oven to 180°C/gas 4. Remove the cores from the tomatoes (if large) and cut the flesh into rough cubes of about 2cm (if using cherry tomatoes, just cut them in half). Set them aside.

Dry the cod skin and season on both sides. Place a large non-stick ovenproof frying pan over a medium heat, add the oil and wait until it just starts to smoke, then place the cod, skin-side down, in the oil and leave it until the skin is lightly golden brown. Transfer the pan to the oven and continue to cook for about 8–10 minutes, depending on the thickness of your cod – it should be just opaque.

While the cod is in the oven, heat the grill or a griddle pan. Cut 2 of the lemons in half and grill them (cut-side up) or griddle (cut side down) until charred.

Remove the cod from the oven and transfer to four warm plates. Place a pan on the hob and add the butter. Season with a little salt and pepper, and when the butter starts to foam and turn a golden nut-brown colour, add the tomatoes and cook for 2 minutes, to let them soften slightly. Add the juice of the remaining 2 lemons, capers and parsley and spoon over the cod. Serve with the grilled lemon halves and the watercress, if using.

SERVES 4

Halibut with Morecambe Bay Shrimp Butter Sauce

Tiny, brown, quite sweet-tasting shrimps have been caught in Morecambe Bay on the Lancashire coast since the eighteenth century. They are boiled up straightaway, then peeled. Most of them go to be potted in the traditional way (cooked, then sealed in butter with a little warm spice), but you can also buy them peeled to cook yourself. In this dish, we keep to the spirit of potted shrimps, adding them to a butter sauce, flavoured with nutmeg and herbs to serve with the halibut. We source Scottish farmed halibut as it is the most sustainable option from British waters.

100g butter

1 large shallot, finely chopped

1 clove of garlic, finely chopped

200ml white wine

2 tablespoons double cream

juice of ½ a lemon

100g peeled Morecambe Bay brown shrimps

½ teaspoon ground nutmeg

1 tablespoon chopped fresh chives

1 tablespoon chopped fresh chervil

1 tablespoon chopped fresh flat-leaf parsley

400g purple sprouting broccoli, stalks removed

2 tablespoons olive oil

4 x 200g halibut steaks

sea salt and freshly ground black pepper

Melt 1 tablespoon of the butter in a small pan, add the shallot and garlic and cook over a low heat for 10 minutes, until the shallot has softened but not coloured. Add the white wine, bring to a simmer and let the liquid reduce by two-thirds. Add the cream and bring back to a gentle simmer, then reduce to low. Whisking swiftly, add all but a knob of the butter a little at a time, making sure the sauce doesn't boil or the butter will split and the sauce will curdle. Once the butter is added and the sauce is thick and silky, add half the lemon juice, with the shrimps, nutmeg and herbs. Take off the heat and keep warm.

Meanwhile, steam the purple sprouting broccoli until just tender (usually 3–4 minutes, depending on how crunchy you like it).

Heat the olive oil in a large non-stick frying pan over a medium heat. Season the halibut and put it into the pan. Cook for 3–4 minutes on each side, until golden brown and just opaque in the centre. Squeeze the rest of the lemon juice over the fish, add the knob of butter and spoon over the juices from the pan. Arrange the halibut on warmed plates with the broccoli. Spoon the shrimp sauce over and serve.

SERVES 4

Clam Linguine

A classic Italian *pasta alle vongole* is one of my favourite dishes – it is the essence of summer and conjures happy memories of holidays on the Italian coast. John tells me that the choice of wine is key – you want something not too fruity, a little drier, with a good edge of acidity and sharpness, like a Sauvignon Blanc or Muscadet.

1.5kg fresh clams, in their shells

200g butter

2 small white onions, finely chopped

4 cloves of garlic, finely chopped

1 fresh red chilli, very thinly sliced

400ml white wine

200ml double cream

400g good-quality linguine (fresh or dried)

juice of 1 lemon

1 small bunch of fresh tarragon, chopped

1 small bunch of fresh flat-leaf parsley, chopped

sea salt and freshly ground black pepper

Wash the clam shells well under cold running water and discard any that are open or that won't close if you tap them.

In a large wide pan, melt 2 tablespoons of the butter. Add the onions and garlic and cook over a low heat for 10 minutes, until the onion is soft but not coloured. Add the chilli, clams and white wine and bring to a rapid boil. As soon as the clams open, lift them out of the pan with a slotted spoon and put them into a bowl. Remove the shells from half of them and discard these – so you have a mix of clams in and out of shells. Bring the liquid in the pan to a rapid boil to allow it to reduce by half, then add the cream and bring back to a simmer.

Meanwhile, drop the linguine into a large pan of salted boiling water. Cook until just al dente (if using fresh pasta, this will probably be less than 2 minutes). Drain, then add to the pan of sauce, along with the reserved clams.

Toss the pasta lightly with the sauce, adding the remaining butter a knob at a time. When it is all absorbed and the sauce is thick and silky, add the lemon juice and herbs. Taste and season as necessary, and serve immediately.

SERVES 4

Baked Salmon, Spinach and Smoked Haddock Kedgeree

This is a bit of a twist on the more traditional kedgeree – the chefs cook the fish in its spicy, creamy sauce and serve it separately to the rice so that you can mix everything together yourself, having more or less rice, as you prefer.

4 eggs

300ml good vegetable stock

1 large (300g) fillet of smoked haddock, skinned and boned

400g brown rice

2 tablespoons butter

1 medium onion, finely sliced

1 clove of garlic, finely chopped

4 tablespoons medium curry powder

1 tablespoon onion seeds

2 tablespoons finely chopped fresh ginger

1 large medium-hot red chilli, seeds removed, very thinly sliced

200ml double cream

4 tablespoons natural yoghurt

2 tablespoons chopped fresh parsley

2 tablespoons chopped fresh coriander

1 handful of spinach leaves

1 x 350g fillet of salmon, skinned and boned, and cut into bite-size pieces

2 tablespoons dried breadcrumbs

sea salt and freshly ground black pepper

Have the eggs at room temperature. Bring a medium pan of water to the boil and add a pinch of sea salt (this will make the eggs easier to peel). Gently lower in the eggs and simmer for exactly 7 minutes. Take off the heat and rinse them immediately in cold water to prevent discolouring, then peel the eggs.

Preheat the oven to 180°C/gas 4.

Pour the stock into a pan, bring to a simmer and drop in the fillet of haddock. Simmer for 5 minutes, then remove the pan from the heat and allow to cool slightly. Lift out the haddock and flake it, then strain the stock into a bowl and keep to one side.

Put the brown rice into a large pan with plenty of cold salted water, bring to a simmer, and cook, covered, for about 25–30 minutes, until it is soft and tender. Strain through a colander, then put the rice back into the pan with a knob of butter, season it, stir, put the lid back on and leave it to steam (off the heat) until you need it.

Meanwhile, melt the remaining butter in a large pan (one that has a lid) and add the onion and garlic. Cook over a medium heat for 5 minutes, until the onion is soft, but not coloured. Add the curry powder, onion seeds and ginger, cover and continue to cook for a further 10 minutes, stirring occasionally. Add the chilli and reserved stock, then bring to a simmer and cook until the liquid has reduced down and thickened enough to coat the back of a spoon.

Add the cream, yoghurt and flaked haddock, bring back to a simmer again, then add the parsley and coriander. Taste and season as necessary.

Lay the spinach leaves in an ovenproof dish. Spoon 4 tablespoons of the haddock in its sauce on top.

Cut each boiled egg into quarters and arrange on top, with the pieces of salmon in between. Season, then spoon the remaining sauce over the top. Finish by sprinkling over the breadcrumbs.

Put into the oven for 25 minutes, until the sauce is bubbling and the top is golden brown – but make sure you don't overcook it, or the salmon will become dry. Serve with the brown rice.

The Animals
Richard Smith

I ache all over to try to get people to understand that farming can't carry on the way it is going. Intensive monocultures don't work; agriculture has to have diversity in order to be sustainable. Yes, we have more and more mouths to feed, but the way forward is to eat a little less meat, but of better quality, produced locally, from animals that are raised slowly and naturally on grass, in sustainable systems, without the artificial fertilisers and pesticides that release greenhouse gases, so the environment is being nurtured, rather than destroyed.

The way I feel about farming has evolved over many, many years, since I first knew – without question – that I wanted to be a farmer, when I was ten years old. I've seen and worked in lots of different systems, but I think British agriculture at its finest is the most diverse – and the best in the world.

There are a lot of farmers out there, who say, 'Bloody organic . . . can't do this, can't do that.' That's rubbish. The reality is that we are only two generations on from a time when most farming was organic anyway. My view is that if you go back to the absolute basics of agriculture – raising animal welfare standards, concentrating on the right breeds for the environment you want to raise them in, managing the production of great grass for grazing, great crops for silage, great hay and the conservation of feed for winter – then the end result is you can produce food of superb quality, without chemicals in the soil and routine medicine for the animals.

I prefer the word sustainable to organic, which has been hijacked somewhat. What organic certification gives people is an assurance of specific standards, levels of animal welfare and traceability. Much of the food and farming industry has carried on behind closed doors and people haven't always known what they were buying; haven't even really wanted to think about it. Cheap food has been far too easy to come by and good food hasn't been respected enough, which is why we waste so much. But now more and more people do want to know, and do care. What we stand for here at Daylesford is the honest and transparent production of food, and we are more than happy for people to visit us and see what we do.

We have a lot of requests from students and also farming groups to come and look around. Often their background is in quite intensive systems and I can see in their eyes that they are expecting me to be a tree-hugging, woolly, homespun sort of fellow, and they are going to see a less efficient, less productive standard of agriculture. I put them on a trailer and take them round the farm, talk to them about crop rotations, properly costed-out systems and show them our flocks and herds that are not reliant on cereals and high-energy proteins forced down their throats, but thrive on a home-grown, diverse, forage-based diet. Suddenly people are listening intently because they realise we really are on to something and they mob me for information.

I have always lived on a farm. My dad worked on a large mixed farming estate in Northamptonshire and, as soon as I was old enough, I was pestering him and the neighbouring farmers to let me help out. My biggest farming role model was my grandfather, a proper no-nonsense, incredibly knowledgeable character from Northumberland, who came to the Midlands to work as a shepherd in the 1950s.

My uncle is also a tenant farmer on a hill farm in the Cheviot Hills in Northumberland, a very passionate, specialist sheep breeder, so after I left school I went to live with him for six months of each year for five years, then came home for the harvest, hay-making and autumn cultivations. I loved all of it, but what really drove me was the livestock.

When I was 24, I spent two years managing a farm in Cornwall, specialising in beef cattle and sheep – I loved the place and the people, but it was very intensive and fertilisers were used heavily. Afterwards I worked briefly in Kent, managing one of the largest beef herds in the country and I can honestly say it was the most unhappy I have ever been in my farming life: I hated the whole intensive system. So it was with great relief that, after I met and married my wife, Claire, who came from New Zealand, and we moved there for seven years, I was lucky enough to be offered a share-farming agreement near Wellington. It was a huge break for me and a very special time. I farmed about 2,500 local breed, grass-fed Romney sheep, venison and Aberdeen Angus cattle.

When Claire died suddenly, I came back to England in 1999 and was fortunate enough to be offered a job at the Oxford University Farm, where they research and develop sustainable farming systems built around animal welfare and care of the environment – not far from Daylesford. So, if you believe in divine intervention, I was destined to meet Carole Bamford in 2005. What excited me was that she very genuinely wanted to do something to shout about in agriculture. I knew instantly that this was where I wanted to be – what an opportunity we had to raise the bar for British farming, to help to make organic, sustainable agriculture credible and respected. Eight years on, we now farm 3,4800 acres in Gloucestershire and a further 3,000 on the Wootton Estate in Staffordshire, and I want us to be as self-sustaining as possible, so we are constantly planting new acres of silage crops and pastures for grazing. Sheep and cattle are ruminants that have evolved to make the most of forage-based diets, to walk around the countryside, eating different things at different times, and their stomachs are designed for that, so the more diverse and species-rich you can make those pastures, with a mixture of grasses, herbs, legumes and clovers, the better. Something will be happening all year round: plants flowering, coming into leaf, forming seed heads. . . and foraging on them is what helps to keep an animal healthy. Stockmen have always understood that – that is what the system of transhumance is all about in mountainous regions of Europe, moving herds up and down the mountains and from pasture to pasture, according to season, so that the animals have variation in their diet.

Cotswold sheep

Sustainable farming depends on the right breeds of animal in the right place, and over the years we have been building our own carefully selected closed herds, so that every generation is more suited to the environment I am asking it to live in. Take sheep: I could talk all day about farming sheep. We have breeds in Britain that have evolved and been cross-bred over hundreds of years to suit the particular terrain and environment in different parts of this island that we live on. You can go all over Europe and see genuine old breeds of animals, but they are not performing in the way ours do. The Cotswolds, in particular, were built on sheep: the old English word 'cot' refers to the compounds where sheep are kept, and 'wold' means rolling hills, and so it seems only natural that we have a flock of native Cotswold sheep – one of four heritage breeds we now have on the farm.

If you take two animals of pure blood and cross them, the progeny they produce will inherit traits from both the maternal and paternal lines. This is what is known as 'hybrid vigour'. For instance, generations of farmers in the north of England and the Scottish Borders have specialised in breeding sheep that are wonderfully adapted to the local terrain, crossing ewes from naturally hardy hill breeds, such as Scottish Black Face and Swaledale, with more prolific, faster-growing animals, such as Border Leicesters and Blue Face Leicesters, and their progeny are known as 'Mules'. The hardy hill breeds have coarse fibrous wool, so they can survive tough winters, and a metabolism that helps them run on brackens and poor-quality grasses up on the hills; while the less hardy, high-performance breeds, with their fine wool and Roman noses, need easier, higher-quality grazing – but bring the two together and you have a sheep that has the hardiness of its mother and also the growth rate, prolificacy and improved conformation of its father.

We breed our own Lleyn sheep. They are a Welsh breed developed on the Lleyn peninsula in North Wales, about two thirds of the size of a north-country Mule, but they produce a wonderful amount of meat. By closing off the flocks in this way, and keeping the bloodlines as pure as possible, we have complete control of the breeding and health status of our flocks.

We also have a flock of Ryelands, one of the oldest of British breeds – apparently Queen Elizabeth I insisted on her stockings being made only from Ryeland wool – and a pedigree flock of Kerry Hills, a very old breed from the English-Welsh borders. They are very striking-looking – the lambs in particular have endearingly funny faces, with black noses, panda eyes and pricked black ears, but they produce beautiful, succulent, full-flavoured meat.

The skill in rearing grass and forage crop lamb is to consistently produce the same size of carcass and quality of meat 52 weeks of the year and to know when an animal is at its prime and ready for slaughter. When you have graded as many lambs as I have you can tell just by looking at them and running your hands over their backs.

People go mad for 'spring lamb' for Easter, but grass-fed lambs are naturally *born* at Easter – so young lamb that is for sale at Easter comes from animals that were born in November and reared on concentrated diets. I don't want my lambs born when there is likely to be snow on the ground and they may never go outside. All our lambing is in spring and the animals graze on pasture throughout the summer, so the lamb we sell at Easter comes from hoggets – lambs that are born late in May and are nearly a year old.

Animal welfare

One of the questions visitors to the farm usually ask, is 'What do you do if you have a sick animal?' Well, open any guide to organic farming and animal health, and welfare is at the top of the agenda. Of course if an animal is unwell, you treat it, but you simply have to record everything carefully, and withdraw the milk or meat for a specific period of time. The welfare of our animals is a huge priority here, even if that means that we produce less meat or milk or eggs than in intensive systems. Our chickens and hens forage where they like, as do our traditional bronze turkeys, which are funny things: on a bright sunny day they won't come out of their houses, because they think a predator is going to dive out of the sky and attack them, unless they have trees to range under. So we planted trees all around their pastures, which they rush to when the sun comes out, and then they are happy because they feel protected.

At Wootton, we rear chickens for the table. One of the challenges facing so many farmers is that chickens are at risk of developing health problems. Newborn chicks are very fragile and the first few hours of their life are crucial. We are very proud of having our own hatchery where our team is able to monitor the chicks after they hatch to guarantee the safest conditions. From there they go into our brooder sheds, then out on to 100 acres of grassland. The rearing pocess is completely transparent, with animal welfare at its heart, and it also allows us to produce the very best quality of meat.

Probably the animals that live closest to the way they would in the wild are the deer, which are all farmed at Wootton, where they roam and graze hundreds of acres of parkland. There have been herds there for thousands of years, but they have only been farmed seriously since the 1980s. Organic deer farming is all very natural and the animals are true foragers with a real free spirit. Our deer are tended to by just two stockmen, who have a deep knowledge and understanding of their characteristics and how they behave. They can live for twelve to sixteen years and have a fantastic life.

We also want to ensure that the end of the animals' lives is as peaceful and stress-free as possible, which is why we built our own abattoir at Wootton. The animals are settled for at least a day in the fields before going into the straw-bed barn of the abattoir. There is a vet present at all times and it is as humane a process as is possible.

Dairy cows

One of the animals that can really suffer from stress in intensive systems is the dairy cow. We have Friesians, fed on a home-grown diet, and they produce an average of 7,000 litres per lactation, in comparison to the 10–12,000 litres that you would expect from the Holstein breed – which is the dairy equivalent of a Formula One racing car. The vast majority of milk in this country, throughout Europe and large areas of the world, comes from Holstein cattle – but you can't put a racing car on a country lane, and likewise, you can't let a high-producing Holstein milk machine just forage on grass. It has to be fed around 4 tonnes of high-energy, compound feed. The higher the protein, the higher the fuel and the higher you are going to get your Formula One cow to perform. Farmers have been pushed into this corner because they have had to become more and more efficient, and are expected to get more from each animal and every acre of land. Often the result is that the cow's metabolism is stressed and its reproductive system upset – and if a cow can't reproduce, then it can't calve, and enter a new lactation.

Thirty-forty years ago, the British dairy industry was the backbone of the British beef industry because the male calves, which obviously couldn't produce milk, would be reared for their meat. It's a well-known fact in farming that beef from dairy animals raised on grass is always good to eat, but these days most farmers can't afford to keep them for beef, as ultimately they don't yield enough meat to make them profitable in comparison with pure beef cattle.

In the big dairies, if a calf is a female it will be reared for the milk chain, but if it is male, it is often considered to be of no further use, and statistics show that at least 100,000 male calves are shot at birth every year. Here, though, our male Friesian calves are raised on grass, either for rose veal or we allow them to grow to become mature beef.

Our Gloucester herd

In 2006 we established our herd of Gloucester cattle. The Gloucester breed can be traced back to the thirteenth century – they are one of the original 'dual purpose' breeds and a family would have kept one in the byre to provide them with both milk and beef. Until recently there were very few left in the world, but we stumbled upon a herd that was in poor health and built it up with the help of a beautiful champion cow with a lovely nature called Peglards Lola and a handsome bull called Ambrosia Hethelpit Cross. Now we have seventy registered animals, which I believe is the biggest herd in the world. The Gloucesters have become a minority breed in terms of dairy farming for a reason: they can't keep up with the pressures of modern expectations and don't give a huge quantity of milk, but what you do get is of lovely quality. And having the herd means that we can also make Single Gloucester cheese. Above all, they produce meat of superb quality: the marbling, or fat content, texture and taste is just beautiful, and so every so often we are able to release a limited amount of it to the farmshop. We once invited people to taste the Gloucester beef. Before they tucked in I told everyone, 'What you have in front of you is part of our living heritage, 700 years in the making, and no else else in the world will be eating what you are eating tonight.' Producing food that special is hugely satisfying.

Heritage breeds

Carole Bamford is a natural 'stockman', with an eye for classy stock – it was she who researched and first suggested adding a herd of South Devon cattle to the Gloucesters and the Aberdeen Angus. Another old breed, they are also known as Orange Elephants because of the colour of their coats and their size: a full-grown bull can weigh one and a half tonnes. They are magnificent, strong, but gentle creatures, and were well known in England in the eighteenth and nineteenth centuries, when they had a triple purpose: producing milk, beef and working as draught animals that would have pulled ploughs and carts.

I had initially favoured a herd of Herefords, but Carole and I went to see both breeds at a show and there was a butcher there, so she bought a rib joint and sirloin steaks from each to take back to the farm for a taste test. While the butcher was wrapping the meat, Carole asked him which breed he thought had the best flavour and quality. I was standing behind her mouthing: 'Hereford,' but he told her: 'South Devon.' We cooked the meat and tasted it, and everyone – including me – agreed that the South Devons were the right choice. The idea is that we will grow the herd and cross a percentage of them with our Aberdeen Angus herd, which has been in the Bamford family for many years, but which is now 100 per cent pedigree. The way forward, as with all our livestock, is to continue to breed animals that, through the generations, become more and more suited to this land on which we farm.

MEAT

Notes on cooking meat

The recipes in this chapter focus on beef, lamb, venison and chicken as these are the meats we produce on the farm. The dishes make good use of different cuts, from prime ones for roasting, to the cheaper cuts, such as shin and shoulder, which are full of flavour, but need long, slow cooking.

A tip I had from the Daylesford chefs is that when you buy beef, look for meat that has been hung and will look a rich ruby red colour, rather than crimson. There is a tendency to see bright red meat and think it looks fresher, but meat that hasn't been hung will be less mature in flavour and likely to be less tender when cooked. The purpose of hanging beef is to relax it and concentrate the depth of its flavour – ours is hung for 28 days. Lamb, too, benefits from hanging, but for a shorter time: just a week or two, while our venison is only hung for a week, otherwise its flavour can become too gamey.

Steak is always a very popular cut on our butchers' counters, but I know the cooking of it can be tricky, and if you're going to invest in a good piece of organic meat, you want to make sure that you cook it to perfection. I spoke to John Hardwick, Daylesford's Director of Food – a brilliant chef – and he offered me his tips on cooking steak to ensure it remains succulent and juicy.

'First of all,' says John. 'Take it out of the fridge for about an hour before you want to cook it, to allow it to relax. Every chef has their own opinion on whether to season the meat before or after it goes in the pan. The one thing everyone agrees on, though, is that you don't want to add salt too early or it will draw the moisture out of the meat, making it tougher. I believe in adding plenty of salt and pepper just before the meat goes into the pan, so that the seasoning cooks into it.

'You can barbecue, griddle or grill a steak, but I like to cook it in a pan in butter and oil – one-third oil to two-thirds butter. The oil is only there to stop the butter burning.

'Choose a vegetable oil or an ordinary olive oil – not extra virgin. All oils have different smoking points, which means they break down at different stages of heat and their character and goodness can alter. Most oils that have high smoking points also have a neutral flavour. Rapeseed oil is the one that bucks the trend as it has one of the highest smoking points, but also a very strong flavour – which people either like or don't. Personally I think a more neutral oil is better for cooking steak as it won't compete with the flavour of the meat.

'Heat the oil in your pan first, without the butter, until it is on the verge of smoking. Season your steaks on both sides and add them to the pan. Cook until they start to colour slightly, then add the butter. The temperature at which you cook your steak is very important: you need to get a good colour and caramelisation on the outside, which really enhances the flavour, but be careful not to burn the meat. What you are listening for is a nice gentle sizzling sound as the steak cooks. As the butter foams, spoon it over the top from time to time. Give each steak 2–3 minutes without moving it, until it seals and turns golden brown underneath, then turn it over.

'I think a steak should always be served medium–rare, so that the fat can break down and flavour the meat during cooking. How long it takes to get it to this stage depends on the size and thickness of your meat, but when you cook steaks regularly, you learn to gauge it by touch. If you press the meat with your thumb before it goes into the pan, you will get the feel of it when it is raw: it will be very firm. As it cooks, keep testing with your thumb and, when it gets to the rare stage, it will be really bouncy. If you take it off the heat at this point, lift it out and let it rest, by the time the heat has spread through the meat as it relaxes, it will be a perfect medium–rare. If you let the steak continue to cook and keep pressing it, you will feel that it gradually has less and less give, until it has no give at all, at which point it will be well done. Once you have tried this a few times, you will be able to recognise the stages and tell just by the feel of it when the steak is done to your liking.

'There is another pretty foolproof way to tell when the steak will be medium–rare, which works for roast meat too (see page 224). Run a cold tap over a metal skewer, so that it is very cold, insert it into the centre of the steak, leave it there for 5 seconds, quickly remove it and put it carefully on the back of your hand. If the skewer still feels cold, the meat isn't ready. If the chill has just gone off the skewer, then it is just right; and if it is about to burn your hand, it is overcooked.

'Finally, it is really important to let the meat rest. The rule is to rest it for as long as you cook it. If you serve the steak straight from the pan, with no resting, when you cut into it you will see the outside looking quite caramelised and brown, and underneath it a line of greyish cooked meat, which will give way to pinkness, and then in the centre, it will be red-raw. If you let it rest, however, it gives the fibres of the meat time to relax and it allows the heat to carry on penetrating, so that the meat is uniformly tender and juicy all the way through.'

SERVES 6

Lamb and Butter Bean Casserole with Tomatoes, Caperberries and Olives

This is a lovely warming winter dish, especially served with mashed potatoes. You can use good jarred or tinned beans, if you are pushed for time, but our chefs always cook butter beans from scratch, as they absorb the juices and flavours in the casserole better and they are very simple to do. You just need to soak the dried beans overnight first before cooking them separately from the casserole, only combining the two at the end, otherwise you can compromise either the meat or the beans, if one or the other isn't ready.

If you want to cook your own beans, as a rule of thumb, all pulses (and rice) will double in size once cooked. So if a recipe calls for 450g of cooked butter beans, as this one does, start with 225g of dried beans. Put them into a bowl and add enough cold water to cover them by about a centimetre. Leave to soak overnight, then drain the beans, rinse, transfer to a large pan, and again, add enough cold water to cover by a centimetre. Don't add any salt as this will harden the skins of the beans. The key to cooking any pulses is not to boil them rapidly (except for the initial cooking of kidney beans) as they will break up, so bring them to a simmer, then turn the heat right down, so that there is virtually no movement in the water, and cook them for 45 minutes to 1 hour, until the beans are perfectly soft and creamy, but still intact.

1.5kg lamb neck fillet, cut into cubes of around 4cm

3 teaspoons ground coriander

100ml olive oil

2 red onions, thinly sliced

½ a red chilli, finely chopped

1 heaped tablespoon tomato purée

250ml white wine

1 litre good chicken stock

450g cooked butter beans (see introduction, above)

zest of 1 large lemon

1 clove of garlic, thinly sliced

300g cherry tomatoes

50g black Kalamata olives, pitted

70g caperberries, stalks removed, halved

2 tablespoons chopped fresh flat-leaf parsley

sea salt and freshly ground black pepper

Lamb and Butter Bean Casserole continued

Preheat the oven to 160°C/gas 3.

Season the lamb with salt, pepper and the coriander. Heat the oil in an ovenproof casserole (one that has a lid), then add the lamb and brown it all over.

Lower the heat and add the onions, chilli and tomato purée. Continue to cook for 2–3 minutes, until the onions have softened slightly, then add the white wine and bubble up until reduced by half.

Add the stock and bring to the boil, then put the lid on the casserole and transfer it to the oven.

Cook for about 2 hours, until the meat is soft and tender and the gravy has thickened, then take the casserole out of the oven, put it back on the hob and bring to a simmer. If you need to thicken the gravy a little, let it bubble up and reduce for a few minutes before adding the beans. If not, add them straightaway, along with the lemon zest, garlic, tomatoes, olives, caperberries and parsley. Bring back to a simmer, then take off the heat and serve.

Slow-cooked Lamb Shoulder with White Beans and Salsa Verde Mayonnaise

This is a wonderful slow-cooked and warming stew that really showcases the lamb. It does involve a little time, as the meat needs to go into the oven first, so that the fat melts through it and it becomes incredibly tender – but it's well worth it.

4 tablespoons olive oil, plus a little extra to finish

6 cloves of garlic, chopped

1 teaspoon fresh thyme leaves

1.5kg shoulder of lamb, boned and rolled

2 carrots, roughly chopped

1 large onion, chopped

1 leek, chopped

1 heaped tablespoon tomato purée

125ml red wine

1 litre good chicken stock

salsa verde mayonnaise (see page 336), to serve

sea salt and freshly ground black pepper

For the beans:

150g dried white beans

25g butter

1 small onion, finely chopped

1 clove of garlic, crushed

1 carrot, finely chopped (about 5mm)

2 tablespoons chopped fresh flat-leaf parsley

1 tablespoon chopped fresh tarragon

1 tablespoon chopped fresh mint

150g baby spinach leaves, roughly shredded

Mix 2 tablespoons of the oil with the garlic, thyme and salt and pepper and rub all over the lamb. Leave to marinate overnight in the fridge. Soak the beans in cold water overnight, too, then drain.

The next day, preheat the oven to 160°C/gas 3.

Heat the rest of the oil in a casserole (one that has a lid), put in the lamb and brown on all sides, then lift out and keep to one side. Add the carrots, onion and leek and cook over a low heat for 5 minutes, until the vegetables have softened but are not coloured.

Add the tomato purée and the red wine and bubble up to reduce the liquid by half. Return the lamb to the casserole, add the stock, and pour in enough water to cover the meat. Bring to a simmer, cover and put into the oven for about 3 hours, until the lamb is very tender.

Slow-cooked Lamb Shoulder continued

Lift out the lamb from the casserole and keep to one side (reserving the cooking liquid).

To cook the beans, heat the butter in a large heavy-based saucepan. When it's foaming, add the onion and garlic and cook gently for 5 minutes or until the onion is translucent. Add the carrot and the drained white beans and strain in the reserved cooking liquid from the lamb. Bring to the boil, then turn down to a simmer for 30–40 minutes, stirring occasionally, until the beans are tender (you may need to top up with a little water).

Remove from the heat and stir in the herbs and spinach. Taste and season as necessary.

Return the cooked lamb to the pan over a low heat until heated through.

To serve, lift out the lamb and slice into 4. Spoon the beans and sauce into a warmed, shallow serving dish or platter. Arrange the lamb on top and drizzle with a little olive oil. Serve with a bowl of salsa verde mayonnaise.

SERVES 4

Pressed Lamb

A great thing to do when you have invited friends round is to slow-cook the lamb as in the previous recipe, but then press, slice and pan-fry it, so that the slices are crispy on the outside and meltingly soft in the middle – it is really beautiful this way. You do all the major cooking in advance, then shred the meat with some of the sauce and form it into a sausage shape, which can be chilling in the fridge while you relax and enjoy yourself. Then you only have to slice the lamb, pan-fry and serve it. It goes especially well with smashed broad beans and peas (see page 121), which can be made really quickly while the lamb slices are in the pan, and the minted aioli (see page 340).

You will probably have a little more sauce than you need, so any left over can be kept in the fridge for 4–5 days or in the freezer. If you are making a Sunday roast, just add it to your pan of meat juices and loosen with some boiling water to make a great gravy.

To make the pressed lamb, follow the previous recipe, but omit the beans. After the lamb has been in the oven for about 3 hours, or is very tender, lift out the meat with a slotted spoon and leave it to cool.

Strain the cooking liquid through a fine sieve into a medium pan, bubble it up and leave it to reduce it to a thick sauce, the consistency of ketchup, then take off the heat.

When the lamb has cooled, flake it into a bowl, discarding the fat and gristle, and stir in 4 tablespoons of the reduced sauce. Taste the mixture and season as necessary, then stir in a tablespoon each of chopped mint, tarragon and parsley.

Cut 2 squares of clingfilm, about 30cm x 30cm, and lay one on top of the other. Spoon the mixture on to the clingfilm, form into a cylinder shape (about 7–8cm in diameter), then roll up tightly, twisting and tying the ends. Wrap a layer of foil around the outside and again twist the ends, so that the roll looks like a Christmas cracker.

Put into the fridge to chill for at least 2 hours. When ready to serve, cut the lamb into 8 slices. Heat a little olive oil in a pan and fry the slices until they are golden on each side and hot in the middle. Serve with smashed broad beans and peas, and minted aioli.

SERVES 4

Chicken Casserole with a Splash of Brandy

This is a version of the Belgian dish, chicken Waterzooi, which my children adored when they were growing up. It is lovely and light, yet comforting – there is nothing like it, especially when you're tired. The casserole is good served with simple boiled potatoes.

90g butter

1 free-range chicken

2 tablespoons brandy

1 bay leaf

500ml good chicken stock

2 carrots, sliced

2 large leeks, white part only, sliced

4 sticks of celery, sliced

250g small mushrooms, sliced

4 tablespoons chopped fresh flat-leaf parsley

3 egg yolks, beaten

100ml double cream

sea salt and freshly ground black pepper

Preheat the oven to 160°C/gas 3.

Heat a third of the butter in a large flameproof casserole (one that has a lid). Season the chicken and, when the butter is foaming, put it into the casserole and brown on all sides. Add the brandy and allow the alcohol to burn off, then put in the bay leaf, stock and enough water to just cover the chicken and bring to the boil.

In another pan, heat half the remaining butter and put in the carrots, leeks and celery. Season, sauté until golden brown, then add to the casserole.

Heat the remaining butter in the same pan and put in the mushrooms. Season and, once again, sauté until golden brown. Add to the casserole with half the parsley, cover with the lid, and put into the oven for 1 hour, or until the chicken is cooked (pierce the thigh with a skewer and the juices should run clear).

Lift out the chicken and put it on a board for carving. Lift out the vegetables with a slotted spoon and transfer to a large, warm, shallow serving dish.

Chicken Casserole continued

Strain the cooking liquid into a bowl, then pour it back into the casserole and bubble up on the hob until it has reduced by three-quarters and is thick enough to coat the back of a spoon.

While the sauce is reducing, joint and carve the chicken and arrange the pieces on top of the vegetables.

Turn off the heat under the sauce and stir in the remaining parsley, egg yolks and cream, making sure the sauce doesn't boil or the eggs will scramble. Pour over the chicken and vegetables, and serve immediately.

SERVES 6

Chicken and Apricot Curry

The squash and apricots give this autumnal curry a gentle sweetness. For the best flavour, use a mixture of breast and thigh chicken meat. Our chefs serve this with brown rice.

- 1.2kg chicken meat, cut into bite-size pieces
- about 2 tablespoons sunflower oil
- 1 large onion, finely chopped
- 2.5cm piece of fresh root ginger, finely grated
- 2 cloves of garlic, finely chopped
- 1 red chilli, finely chopped
- 1 teaspoon cumin seeds
- 1 tablespoon curry powder
- 1 tablespoon chilli powder
- 1 tablespoon ground turmeric
- 2 large ripe red tomatoes, roughly chopped
- 150g squash, peeled and chopped
- 1 large baking potato, chopped (about 2cm)
- ½ a medium leek, sliced
- 100g apricots, roughly chopped
- 500ml good chicken stock
- 100ml coconut milk
- sea salt and freshly ground black pepper

Season the chicken. Heat 2 tablespoons of sunflower oil in a large casserole and brown the chicken in batches, removing each batch to a bowl.

Lower the heat and add the onion, ginger, garlic and chilli, together with a little more oil if needed, and cook gently until the onion is softened and golden. Stir in the cumin seeds, curry powder, chilli powder and turmeric and cook for a further 5 minutes, to release the flavour of the spices.

Return the chicken to the pan, together with the tomatoes, squash, potato, leek and apricots, and combine thoroughly.

Add the stock and coconut milk and bring to a simmer. Cover and cook gently for about 30 minutes, until the chicken is cooked through and the vegetables are tender. Taste and season as necessary. Leave to stand for 20 minutes before serving to allow all the flavours to merge, then reheat gently – make sure the chicken is hot all the way through.

Notes on roasts

One of the best ways to showcase good meat is in a classic, timeless roast. However, roasting meat isn't an exact science because every chicken and every joint of beef is different, as is every oven, so it is better to get into the swing of using your eyes, sense of smell and touch to gauge when the meat is ready, rather than relying on charts. Remember, when you are cooking beef or lamb, fat is flavour. An ultra-lean cut will never be as succulent and tasty as a cut that has a fine marbling of fat running through it – which is why when you buy a piece of silverside or topside beef, a butcher will often tie a piece of fat on to it for you.

These are John Hardwick's guidelines:

Rib of Beef: choose a well-hung piece of meat, with a fine marbling of fat. Preheat the oven to 190°C/gas 5. Tie the joint, or have your butcher do this, so that it keeps its shape. If you have kept any beef fat from a previous roasting, heat this in a roasting pan on the hob, or alternatively heat some vegetable oil. Season the beef at the last minute with sea salt and black pepper, put it into the pan and colour it all over, then transfer to the oven and turn the heat down to 180°C/gas 4. As the meat roasts, baste it with the juices and let it cook until the point where, once rested, it will be medium-rare. To test for this, use the skewer test. Run a cold tap over a metal skewer, so that it is very cold, insert it into the centre of the meat, leave it there for 5 seconds, then quickly remove it and put it carefully to the back of your hand. If the skewer still feels cold, the meat isn't ready. If the chill has just gone off the skewer, it is just right, and if it's about to burn your hand, it means that the meat is already well done and becoming dry. Take the meat out and let it rest for about 15–20 minutes, covered with foil, to allow the meat to relax and for the heat to finish transferring through to the centre and, provided it passed the skewer test, it will be perfectly medium-rare.

Lamb Leg: for me, roast lamb leg needs to be medium, not as pink as, say, grilled chops, because the longer cooking allows all the fat to melt into the meat, flavouring and tenderising it. I would cook it in the same way as beef, above, but this time when you do the skewer test, make sure the skewer comes out hot, but not burning, which will mean that the meat will be medium by the time it has rested.

Chicken: remember that organic chickens have a good life, running around in the open air, but that means they develop a lot more muscle than birds that rarely leave their chicken-houses, so their flesh will be firmer and resting will be important after cooking to relax the meat. Start with the oven at 190°C/gas 5. Cut a lemon in half, squeeze some of the juice inside the cavity of the chicken and also season it with sea salt inside. Smear the skin with softened butter or a little olive or vegetable oil, squeeze the rest of the lemon juice over and rub in plenty of sea salt and freshly ground black pepper. The lemon will add to the seasoning and also help keep the chicken moist.

Put the chicken into a roasting tray and put into the oven for 15 minutes, then turn the oven down to 180°C/gas 4 and turn the chicken over so that it is resting breast side down for another 20 minutes. Then turn it back again, so that it is breast side upwards again for another 20 minutes. Now turn the oven back up to 190–200°C/gas 5–6, and cook until the skin has become crisp and golden (this should only take about 10 minutes) and, if you insert the tip of a knife into the thickest part of the thigh, the juices run clear.

Roast Potatoes: I like to roast potatoes around a joint. I boil them first in lightly salted water to the point where they are about to break up, then drain them through a colander over the pan in which they have been cooked, so that they steam and dry out – shake the colander a little to roughen up the edges – then, once your meat is up to speed and sizzling gently, add them to the roasting pan. Don't put them in until the meat has got going or you will bring the temperature down and slow the whole process. Once your meat is out of the oven and resting, turn the oven up to 200°C/gas 6 to get a good golden colour all over – you can add a knob of butter for extra richness, if you like. Once the potatoes start to crisp up, keep turning them over in the oil (it is important to wait until they crisp up, otherwise they will break). If you want to add some slivers of garlic and some rosemary leaves, do this a few minutes before serving, to stop them burning and becoming bitter, and to preserve the green colour and fragrance of the rosemary.

SERVES 2

Roast Rib of Beef with Dijon Mustard and Balsamic Sauce

A rib of beef is perfect for two people. Ours is accompanied by quite a punchy, sweet–sharp mustardy sauce – it's delicious with our potato wedges (see page 126).

- 1 x 700g–1kg beef single rib on the bone
- 1 tablespoon butter
- 100ml extra virgin olive oil
- 1 small onion, finely chopped
- 2 cloves of garlic, finely chopped
- 100ml red wine
- 250ml good chicken stock (see page 373 for how to make your own)
- 2½ tablespoons Dijon mustard
- 1 tablespoon balsamic vinegar
- a little olive oil
- 1 sprig of fresh thyme, leaves only
- sea salt and freshly ground black pepper

To serve:
- 1 small bunch of watercress
- potato wedges (see page 134)

Take the rib of beef out of the fridge to allow it to come to room temperature. Preheat the oven to 180°C/gas 4.

In a small pan, melt the butter and 1 tablespoon of the olive oil. Add the onion and garlic, and cook over a low heat for 10 minutes, stirring often, until the onion has softened but not coloured. Add the wine, then bring to a simmer and let it reduce by three-quarters.

Add the stock, bring back to a simmer and reduce again by three-quarters. Take off the heat and whisk in the mustard and balsamic vinegar. Continue to whisk rapidly whilst very gradually adding the rest of the oil – don't rush this or the sauce won't emulsify properly and will be greasy. Keep by the hob.

Rub the beef with a little olive oil and the thyme leaves, and season with plenty of salt and pepper. Get an ovenproof griddle pan or frying pan smoking hot and put in the beef, cooking it on both sides, until well marked if you are using a griddle pan.

Transfer to the oven for about 15 minutes for medium–rare meat, or longer, depending on how well done you like your beef (see page 224). Remove and leave to stand for 15 minutes, before serving with the warm sauce, watercress and potato wedges.

Featherblade of Beef with Creamed Wild Mushrooms

Featherblade is a great, much-overlooked cut of beef from behind the shoulder and is full of flavour, but because the muscle works very hard, it needs slow cooking. After a couple of hours in a low oven it becomes meltingly tender and is even better if you chill it to firm it up, then slice it into steaks and pan-fry them before serving – then the meat will just fall apart when you put your fork into it. If you want to do this, take the cooked beef from the oven, keeping back the cooking liquor, and roll up the beef tightly in clingfilm. Put it into the fridge for at least 2 hours, to chill and firm up.

When you are ready to serve, take the chilled meat from the fridge and cut it into 6 slices. Heat about 2 tablespoons of olive oil with a couple of good knobs of butter (each about the size of a tablespoon) in a large frying pan and, when the butter is foaming, put in the slices of beef and sauté for 4–5 minutes on each side, until golden brown and hot all the way through. Make the sauce as in the method below, while the slices of beef are sautéing.

Serve with some steamed brassicas, such as kale.

1 beef featherblade (about 1.5kg)

4 cloves of garlic, chopped

1 red onion, chopped

1 stick of celery, chopped

1 carrot, chopped

1 large sprig of fresh thyme

450ml full-bodied red wine, such as Merlot

about 3 tablespoons olive oil

2 tablespoons tomato purée

1 litre good chicken stock (see page 373 for how to make your own)

3 handfuls of mixed wild mushrooms (about 100g), cleaned and stalks removed

5 tablespoons double cream

1 tablespoon chopped fresh flat-leaf parsley

sea salt and freshly ground black pepper

Put the piece of beef into a bowl with the garlic, vegetables, thyme and red wine, and leave in the fridge to marinate for 24 hours.

Preheat the oven to 160°C/gas 3.

Lift the meat out of the marinade, pat dry with kitchen paper and season. Strain the marinade into a bowl, keeping the vegetables.

Heat the olive oil in a casserole (one that has a lid), put in the piece of beef and brown on all sides, then lift out and keep on one side. Put in the reserved vegetables from the marinade, adding a little more oil if necessary, then add the tomato purée and cook for a few more minutes. Add the marinade and let it bubble up, at the same time scraping the caramelised bits from the bottom of the pan. Add the stock, bring to a simmer, then turn off the heat.

Return the beef to the casserole, cover and put into the oven for around 2½ hours, until tender, then remove from the oven. Lift out the beef and leave it to rest, covered with some foil, while you make the sauce.

Strain the cooking liquid from the beef into a medium pan, bring to the boil, then turn down the heat and reduce until you have a sauce-like consistency. Drop in the wild mushrooms, add the cream and simmer for a further minute or two, until the mushrooms are tender. Taste and season as necessary, then finish with the parsley.

Serve a slice of beef on each of six warmed plates and spoon over the sauce. Serve with steamed brassicas.

SERVES 6

Smoky Slow-cooked Shin of Beef Chilli

The touch of coffee in this chilli adds a richness and depth of flavour and colour. You could use tinned kidney beans, but we suggest cooking them from scratch; you start off with 100g and follow the method on page 106. Serve with rice – or even mashed potato.

- 1 teaspoon dried oregano
- 1 teaspoon ground coriander
- 3 teaspoons smoked paprika
- 2 teaspoons hot chilli powder
- 3 tablespoons sunflower oil
- 780g beef shin, trimmed and cut into cubes (about 2cm)
- 40g beef marrow, chopped small
- 2 red onions, each cut into 8 pieces
- 2 cloves of garlic, finely chopped
- 1 red pepper, chopped (about 2cm)
- 600ml good chicken stock (see page 373 for how to make your own)
- 8 red vine tomatoes, roughly chopped (or 600g tinned chopped tomatoes)
- 2 teaspoons sugar
- 6 brown chestnut mushrooms, quartered
- 200g cooked kidney beans
- 1 small green chilli, thinly sliced (keep the seeds)
- juice of 1 lime
- 1 teaspoon instant coffee
- sea salt

Preheat the oven to 150°C/gas 2.

Put the oregano and spices into a dry frying pan over a low flame and toast for a couple of minutes until they release their aroma – take care not to let them burn. Take off the heat.

Heat the sunflower oil in a large casserole (one that has a lid), add the beef and sauté over a high heat until golden brown on all sides, then remove from the pan and keep to one side. Add the beef marrow, along with the onions, season with salt, then lower the heat and cook with the lid on for 5 minutes, until the onions start to soften. Add the garlic and toasted spices, stir and cook for another 2 minutes.

Put back the beef, along with the rest of the ingredients. Bring to a simmer, then transfer to the oven and cook with the lid on for 3 hours. Take out and test a piece of meat – it should be soft and tender, and easily broken with the back of a fork. If it is not tender enough, put it back into the oven for a bit longer. When it passes the tenderness test, taste and season if necessary and serve.

SERVES 6

Braised Brisket with Lentils

Brisket can be a little harder to get hold of than some of the other cheaper beef cuts but it's well worth seeking out or asking your butcher to source it. The key to soft, tender brisket is the thick layer of fat that melts while cooking, so don't trim it.

- 5 tablespoons olive oil
- 1.5kg boned, rolled brisket
- 2 large onions, chopped
- 1 large carrot, chopped
- 2 sticks of celery, chopped
- 3 cloves of garlic, roughly chopped
- ½ a red chilli
- 1 tablespoon tomato purée
- 2 sprigs of fresh thyme
- 4 sprigs of fresh parsley, plus 2 tablespoons chopped fresh parsley to garnish
- 125ml red wine
- 2 litres good chicken stock (see page 373 for how to make your own)
- 1 tablespoon Worcestershire sauce
- 125g Puy lentils
- sea salt and freshly ground black pepper

Preheat the oven to 170°C/gas 3.

Heat the olive oil in a large casserole (one that has a lid). Season the brisket, put it into the pan and sauté gently until golden brown all over, then lift out and keep to one side. Lower the heat, add the onions, carrot, celery and garlic and continue to cook until they are golden brown. Add the chilli, tomato purée and sprigs of thyme and parsley, and continue cooking for a further minute.

Add the red wine and bubble up until the liquid has reduced by half, then put in the brisket, stock and Worcestershire sauce. Bring to the boil, then turn down to a simmer and skim the fat from the surface, cover with a lid and transfer to the oven for 3 hours.

Remove the pan from the oven, lift out the meat and wrap it in foil to keep warm. Strain the cooking liquid into a pan and add the lentils. Bring to the boil, then turn down the heat and simmer until they are tender, but still retain some bite, and the gravy has thickened and coats the back of a spoon. Add the chopped parsley. Cut the brisket into 6 slices, and serve with the lentil gravy spooned over the top.

SERVES 6

Corned Beef

For many people, I know the words 'corned beef' might recall rather unsavoury memories of school dinners. But homemade corned beef is a different affair: it simply involves soaking the meat in a brine before cooking. The salt solution soaks through to the centre of the meat over the course of 8–10 days and gives it its distinctive flavour and pink appearance, which it keeps even after cooking. You can either slice it, to serve it in the traditional fashion with cabbage and potatoes, or allow it to cool and slice it for sandwiches. Alternatively, what our chefs do is flake it, mix it with the thickened cooking liquor, then put it into the fridge so that the natural gelatine in the beef will allow it to thicken and set. Done this way it is good with sourdough bread and piccalilli (or cornichons), or fried up to make our corned beef hash (see page 41).

250g sea salt

60g sugar

1 boned and rolled brisket of beef, about 1.7kg

2 tablespoons olive oil

1 carrot, chopped

2 onions, chopped

1 stick of celery, chopped

1 whole bulb of garlic, sliced horizontally

¼ of a fresh red chilli, deseeded and chopped

1 tablespoon tomato purée

1 small bunch of fresh parsley stalks

1 sprig of fresh thyme

100ml red wine

2 litres good chicken stock (see page 373 for how to make your own)

1 tablespoon Worcestershire sauce

sea salt and freshly ground black pepper

To make the brine, put the salt, sugar and 2 litres of water into a large pan and bring to the boil, then take off the heat and leave to cool. Put the beef into a large casserole (one that has a lid), pour the brine over it, making sure it covers the meat, then cover and put into the fridge for 8–10 days.

Remove the beef from the brine (discarding this) and pat it dry with kitchen paper.

Preheat the oven to 170°C/gas 3.

Wash and dry the casserole, put it back on the hob and heat the olive oil in it. Put in the beef and sear it on all sides until golden brown.

Add the carrot, onions, celery and garlic and continue to colour. When the vegetables are slightly golden, add the chilli, tomato purée, parsley stalks and thyme, and cook for a further minute, then add the red wine. Bring to a simmer and continue to cook until the volume of liquid has reduced by half. Add the chicken stock and Worcestershire sauce and bring to the boil. Turn down to a simmer and skim off any scum from the surface, cover, and put into the oven for 3 hours.

If serving it hot, make an accompanying sauce by straining the cooking liquor through a fine sieve into a clean pan, putting it back on the hob and bringing it to a rapid boil, until it has thickened to a coating consistency.

Alternatively, let the beef cool down in the cooking liquor, then lift it out (but retain the liquor), put the meat into a bowl and flake it into small pieces, removing any gristle, but keeping the fat. Mix vigorously with a fork to allow the pieces of meat and fat to break up. As above, strain the cooking liquor through a fine sieve into a clean pan, put it back on the hob and bring it to a rapid boil, until it has thickened to a coating consistency, then add to the bowl of flaked beef and mix well. Taste and adjust the seasoning as necessary, then transfer the mixture to a loaf tin, smooth the top and put into the fridge for about 2 hours to chill and set.

It will keep for up to a week in the fridge, so you can slice it as required or turn it into corned beef hash.

Beef, Ale and Barley Casserole

SERVES 4

Ale and barley work really well with beef – Daylesford's chefs add the ale in two stages: most of it goes in with the stock, which adds flavour and depth to the casserole, then they add an extra hit of it just before serving, so you get a real, more identifiable sense of the ale.

4 tablespoons sunflower oil

800g stewing beef, cut into cubes (about 2cm)

3 large onions, sliced

2 carrots, sliced

1 medium swede, chopped

2 cloves of garlic, crushed

2 tablespoons tomato purée

2 tablespoons plain flour

1 litre good chicken stock (see page 373 for how to make your own)

90ml pale ale

1 teaspoon chopped fresh thyme leaves

4 tablespoons pearl barley, rinsed in cold water

sea salt and freshly ground black pepper

Preheat the oven to 170°C/gas 3.

Heat the sunflower oil in a large casserole (one that has a lid), put in the cubed beef and sauté over a high heat until golden brown on all sides, then remove from the pan and keep to one side – do this in batches if necessary, so as not to overcrowd the pan and bring the heat down or the beef will steam rather than sauté.

Add the onions, carrots, swede and garlic, lower the heat and cook until lightly coloured. Put in the tomato purée and cook for a few minutes, then sprinkle in the flour. Stir in, then gradually add the stock and two-thirds of the pale ale, stirring constantly to stop the liquid becoming lumpy.

Put back the meat and add the thyme and soaked barley. Bring slowly to the boil, then take off the heat and transfer to the oven. Cook with the lid on for 2–3 hours, until the meat will easily fall apart when pressed with the back of a fork. Stir in the rest of the pale ale, adjust the seasoning and serve.

Venison and Cavolo Nero Lasagne

SERVES 6

This is an autumnal twist on the classic beef dish, which the chefs came up with to make good use of our venison during the game season. Cavolo nero is ready in the market garden at the end of autumn and lasts all through the winter, so it is around at the same time and, by adding a layer of it to the lasagne rather than serving it alongside, you turn the dish into a complete meal. Fresh lasagne is always best, but you could also use the dried sheets that need no pre-soaking.

5 tablespoons light olive oil
1kg minced venison
2 red onions, sliced
3 cloves of garlic, chopped
2 tablespoons tomato purée
300ml red wine
200g tinned chopped tomatoes
1.2 litres good chicken stock (see page 373 for how to make your own)
30g butter
240g cavolo nero, roughly shredded
about 12 sheets of lasagne (see introduction, above)

20g Parmesan cheese, grated
sea salt and freshly ground black pepper

For the béchamel sauce:
1 litre milk
2 cloves
30g onions, finely chopped
125g butter
125g plain flour
200g Cheddar cheese, grated
1 tablespoon English mustard

Heat 2 tablespoons of the oil in a large pan. Season the mince well and brown it in batches until golden, then keep to one side.

Add another 2 tablespoons of the oil to the pan, put in the red onions and garlic and cook gently for 5 minutes until softened, but not coloured.

Add the tomato purée and red wine and bubble up to reduce the liquid by half. Add the tomatoes and return the mince to the pan. Add the stock, bring to the boil, skim off any fat from the surface, then lower the heat and cook over a moderate heat for about 45 minutes. Taste and adjust the seasoning accordingly.

Venison and Cavolo Nero Lasagne continued

Meanwhile, heat the rest of the oil and the butter in a pan, add the cavolo nero and cook briefly over a medium heat just until it wilts. Season well and keep to one side.

Preheat the oven to 180°C/gas 4.

When ready to assemble, make the béchamel sauce by first warming the milk in a small pan with the cloves and onions. Take it off the heat before it reaches a simmer and leave to infuse for 20 minutes. In another pan, melt the butter, then whisk in the flour and cook for 5 minutes over a low heat, stirring constantly. Strain in the milk and continue to cook gently, whisking constantly, for a further 3 minutes until you have a smooth sauce. Stir in the grated cheese and mustard and season well, then take off the heat.

Spoon one-third of the mince into an ovenproof dish (about 25cm x 15cm, and 6cm deep), and cover with a third of the lasagne sheets. Repeat using half the remaining mince and half the remaining lasagne sheets. Finally, spoon in the rest of the mince, followed by the cavolo nero. Cover with the remaining pasta sheets, then pour over the béchamel sauce and sprinkle with the Parmesan cheese.

Put into the oven for about 40–45 minutes, until golden brown and hot throughout.

SERVES 6

Venison Cottage Pie with Beetroot and Apple Salad

This is made and flavoured in the same way as a traditional cottage pie, but is made with venison rather than beef. The autumnal feel is accentuated by serving it with a salad made with apples from the orchard, paired with beetroot. It is also good with just a mixed leaf salad, for something a bit lighter.

Mince, whether it is beef or venison, cooks quite quickly; however, as with any cheap cut of meat, if you want it to become nice and soft and tender, it is best to do it slowly over time, which is why our chefs let the mince simmer gently in its sauce for 2 hours.

2 tablespoons olive oil

1.6kg minced venison

40g butter

1 large onion, chopped

2 large carrots, chopped

4 cloves of garlic, chopped

3 tablespoons tomato purée

500ml red wine

1.5 litres good chicken stock (see page 373 for how to make your own)

sea salt and freshly ground pepper

For the mashed potatoes:

2kg potatoes, peeled and cut into quarters

200ml milk

75g butter

For the beetroot and apple salad:

2 large raw beetroots, peeled and finely shredded

2 apples, cored and finely shredded (with the peel left on)

2 bunches of watercress

2 tablespoons roughly chopped fresh flat-leaf parsley

2 tablespoons sun-dried cranberries, chopped

4 tablespoons French dressing (see page 335)

juice of ½ a lemon

Heat the olive oil in a large casserole. Season the mince, then put into the casserole and sauté until golden brown. Remove from the pan, lower the heat and add the butter, onions, carrots and garlic and cook gently until the vegetables have softened. Add the tomato purée and the red wine and bubble up to reduce the liquid by half.

Return the mince to the pan, together with the stock, bring back to a simmer, and cook over a low heat for about 2 hours, until the sauce has thickened. Take off the heat and leave to cool slightly.

Preheat the oven to 190°C/gas 5.

While the mince is cooking, put the potatoes into a pan and cover with cold, slightly salted water. Bring to the boil, then turn down the heat and simmer for around 20 minutes, until the potatoes are cooked through and easily fall away if pierced with the tip of a sharp knife. Drain in a colander, put into a bowl and mash well (alternatively put through a potato ricer).

Put the milk and butter into a pan and heat until the butter has melted, season and then mix into the mashed potato.

Divide the mince between six individual oven dishes or one large one, then either spoon on the potato and run the prongs of a fork over it, creating little peaks or pipe it, if you prefer. Put into the oven and bake for 20–25 minutes for individual pies or 45 minutes for a large one, until golden brown.

Combine all the salad ingredients in a mixing bowl, gently tossing the dressing and lemon juice throughout. Serve with the cottage pie or pies.

PUDDINGS

Notes on puddings

I have a nostalgic fondness for old-fashioned puddings and our chefs indulge this by creating recipes for traditional puddings that have a contemporary feel, either in the flavours or in the presentation.

Their puddings are never elaborate and they don't deviate too far from the British classics – maybe a twist here and there, or a lighter touch visited on an old favourite, but nothing that detracts from simple, clear flavours. Most of the recipes are built around our own fruit from the gardens, helped along with oranges in winter.

At the height of summer, I like to serve fruit as simply as possible: big bowls of ripe red berries, just as they are, perhaps with a very light sprinkling of sugar and some cream or beautiful juicy peaches, thinly sliced and drizzled with a little sugar syrup (see page 359) and scattered with lemon zest and mint leaves.

MAKES 4

Rhubarb Queen of Puddings

Queen of Puddings is a considered something of a retro classic now, but it makes use of traditional farmshop ingredients: bread, milk, eggs and seasonal fruit, so it's something we've often had on the menu in the cafés. I like it because it looks impressive – the cloud of meringue perched on top of fruit and custard layers is very inviting.

The meringue used in this recipe is Swiss meringue – which John Hardwick insists is the easiest and best to make it. 'French meringue – the more usual kind in which you whisk the egg whites and then fold in the sugar – always has the potential to drop and for the sugar and whites to separate,' he says. 'On the other hand, Italian meringue, while it is more stable, is trickier to make as you have to add boiling sugar syrup to whisked egg whites, and the sugar can stick to the sides of the bowl and not incorporate into the egg whites properly. Whereas Swiss meringue holds as well as Italian meringue and never fails – once you have made it you will never make French or Italian meringue again. You just need a sugar thermometer – a cheap investment that makes working with sugar much easier.'

3 egg yolks
210ml double cream
225ml milk
zest of 1 lemon
30g caster sugar
20g butter
90g fresh white breadcrumbs

For the rhubarb compote:
375g rhubarb, cut into small chunks
zest and juice of 1 orange
175g sugar

For the meringue:
3 egg whites
180g caster sugar
a pinch of salt

Rhubarb Queen of Puddings continued

Preheat the oven to 160°C/gas 3.

Have the egg yolks ready in a bowl. Put the cream, milk, lemon zest and sugar into a pan and heat slowly until the sugar has dissolved, then pour over the egg yolks, stirring constantly. When all is mixed in, stir in the butter and breadcrumbs. Divide this custard between four ovenproof dishes and put these into a roasting pan.

Pour in enough boiling water around the dishes to come halfway up the outsides and put in the oven for 12–15 minutes, or until the custard is just set. Take them out of the oven (but leave it on) and leave the dishes of custard to cool.

Put the rhubarb into an ovenproof dish with the orange zest, juice and sugar. Cover with foil and place in the oven for 30 minutes. Remove the rhubarb with a slotted spoon on to a plate, pour the juice into a pan and bubble up to reduce until very thick, then take off the heat and mix back in the reserved rhubarb. Allow to cool.

When both the custard and the rhubarb are cool, put the egg whites into a heatproof bowl. Add the sugar and salt and place the bowl over a pan of simmering water, whisking until the temperature reaches 60°C on a sugar thermometer, then remove from the heat and whisk until cold, using an electric whisk. The mixture will double in size, form stiff peaks and be smooth and glossy.

Spread some of the rhubarb compote over the custard in the dishes, and spoon (or pipe) the meringue over the top.

Put into the oven for about 10 minutes, until the meringue is straw-coloured.

SERVES 6

Gooseberry Fool with Shortbread

This beautiful fool is rippled, so that you can really see the different layers of gooseberry and creamy custard.

The best gooseberries are usually mid-season, when they are not hard or as acidic as the first ones tend to be, but also not as soft and sugary as the last ones, which will make this pudding too sweet. Try a couple before cooking: what you are looking for is something nice, firm and tart, but not overly so.

400g gooseberries, washed and stalks removed
220g caster sugar
4 egg yolks
60g plain flour
450ml milk

1 vanilla pod, split
350ml double cream

To serve:
6 sprigs of fresh mint (optional)
6 shortbread biscuits

Put the gooseberries into a pan with 100g of the sugar and 2 tablespoons of water and bring slowly to the boil, then turn down the heat and simmer for 5 minutes, or until the liquid has evaporated. Take off the heat and leave to cool.

Whisk the egg yolks, flour and remaining sugar in a bowl.

Put the milk and vanilla pod into a heavy-bottomed pan and heat slowly, taking off the heat just before it comes to a simmer. Add slowly to the egg yolk mixture, stirring constantly, then return to the pan and continue to stir over a low heat until you have quite a thick custard. Strain into a clean bowl and leave to cool, stirring occasionally, then cover with clingfilm to stop a skin forming.

Whip the cream until stiff and gently fold into the cooled custard. Finally fold in the gooseberries very lightly – don't mix them in completely, as you want a 'ripple' effect.

Either serve in bowls or if you want to make the pudding look smarter, spoon carefully into six glasses. Put into the fridge until ready to serve. Garnish each glass with a sprig of mint, if you like, and serve with a shortbread biscuit.

Orange-poached Rhubarb Jelly

I have never lost my childhood love of jellies and will still serve them at dinner parties. Fresh fruit jellies are a lovely, refreshing way to finish a meal and if you fill them with pieces of the fruit, they can look very elegant. The secret to a good jelly is not to use too much gelatine, or you end up with a very bouncy texture.

1.5kg rhubarb, washed and cut into strips

150g caster sugar

zest and juice of 3 oranges

about 3 gelatine leaves

Put the rhubarb into an ovenproof dish with the sugar, orange zest and juice, and leave to marinate in the fridge for 8 hours.

Preheat the oven to 160°C/gas 3.

Cover the dish of rhubarb with foil and put into the oven for 30 minutes, or until the rhubarb is just tender but still holding its shape.

Strain through a fine sieve placed over a bowl to catch the juice. Pour this juice into a measuring jug – you just need 300ml of liquid, so discard any excess. Keep back the rhubarb and arrange it in a serving dish.

Soak the gelatine in ice-cold water (the water must be cold or the gelatine will dissolve) and when soft squeeze out the water.

Pour the rhubarb juice into a pan and heat gently, then add the gelatine and stir until completely melted – but don't boil. Take off the heat, pour over the rhubarb and put into the fridge until set – this will probably take about 3 hours.

Poached Apple and Pear Jelly with Crumble Topping and Prune Cream

SERVES 4

This is an autumnal pudding that our chefs make to use the apples from the orchard at a time of year when there is not too much fruit in the garden to choose from. Setting the apple and prunes in a jelly just makes it a little lighter than a more traditional crumble.

2 small apples, peeled, cored and finely chopped

1 large pear, peeled, cored and finely chopped

375ml apple and prune juice

30g caster sugar

5 gelatine leaves

For the crumble:

100g plain flour

100g butter, chilled and chopped

75g light brown sugar

½ teaspoon ground cinnamon

100g oats

For the prune cream:

75g prunes, pitted

30g caster sugar

½ vanilla pod, split

150ml double cream

To make the jelly, put the apples and pear into a pan with the apple and prune juice, sugar and 100ml of water. Bring to the boil, then turn down the heat to a simmer until the pear is tender. Take off the heat.

Meanwhile, put the gelatine leaves into a bowl of ice-cold water until soft (the water must be cold or the gelatine will dissolve). Take out, squeeze and add to the hot liquid, stirring until dissolved. Divide between four large ramekins or glass jars, allow to cool, then put into the fridge for around 4 hours to set.

Preheat the oven to 180°C/gas 4.

To make the prune cream, put the prunes into a pan with the sugar, vanilla pod and 125ml of water. Bring to a simmer for 5 minutes, then take off the heat and either whizz in a blender or use a hand-held one, until smooth (leave the vanilla pod in, so that it is dispersed through the cream). Leave to cool.

To make the crumble topping, put the flour into a bowl and rub in the butter until the mixture resembles breadcrumbs, then stir in the sugar,

Poached Apple and Pear Jelly continued

cinnamon and oats. Spread the mixture over a baking tray and put into the oven for 12 minutes, until golden brown and toasted.

To finish the prune cream, whip the double cream in a bowl, just until it forms soft peaks, then gently fold into the cooled prune mixture. Do this as lightly as possible so as not to split the cream.

Take the pots of jelly from the fridge and top each one with an equal quantity of the toasted crumble mixture. Serve with a good dollop of prune cream.

MAKES 4

Mulled Wine and Orange Trifle

Our head chef Gaven Fuller came up with this recipe. He had been making mulled wine with oranges and spices to sell in the farmshop, and he thought that the warm, rich flavours would work really well in a jelly for a trifle along with pieces of orange. If you happen to have made the lemon drizzle cake on page 317, a few chunks of it in place of plain sponge cake make the trifle extra interesting.

2 oranges

12 small pieces/cubes of sponge cake or lemon drizzle cake

2 tablespoons brandy

180g double cream

4 tablespoons flaked almonds, lightly toasted in a dry pan

icing sugar for dusting

For the mulled wine jelly:

½ an orange

150ml red wine

a pinch of ground nutmeg

½ a cinnamon stick

a pinch of ground allspice

5 whole cloves

2 tablespoons caster sugar

1 sprig of fresh mint

1 tablespoon brandy

1 tablespoon whisky

3 tablespoons orange juice

2 gelatine leaves

First make the jelly. Peel the skin from the orange and put into a pan. Cut the flesh into rough pieces and add to the pan with the rest of the jelly ingredients, except for the gelatine. Bring to a simmer for 4–5 minutes, then turn off the heat and leave to infuse for 2 hours.

Strain the juice into a bowl through a fine sieve to remove the orange pieces and spices (throw all these away), then return the juice to the pan and warm very gently.

Soak the gelatine leaves in a bowl of ice-cold water until soft (the water must be cold or the gelatine will dissolve), then take out, squeeze and add to the pan of warm juice, stirring until the gelatine dissolves. Take off the heat and keep to one side.

Finely grate the zest from one of the 2 oranges and keep on a saucer (you will be stirring it into the cream later), then peel and segment both. Divide the segments between four tumblers or small bowls and

Mulled Wine and Orange Trifle continued

pour the slightly warm jelly over them. Put into the fridge to set for about 3 hours, until firm.

Arrange the cubes of sponge cake over the top of the jelly and drizzle with the brandy.

Whip the cream with the reserved orange zest until it just falls from a spoon – in kitchens this is called the 'soft flop' – and spoon over the sponge. Put back into the fridge for 30 minutes, then top with the flaked almonds, finish with a dusting of icing sugar and serve.

SERVES 8

Coffee Jelly with Brown Bread Ice Cream

This recipe has been in my family for years and we all still love it. It's something I associate with the start of the children's summer holidays as we'd often make it around that time. It sounds like it is going to be heavy, but it is really light and refreshing, and gives you a lift at the end of a meal.

The jelly contains a lot more gelatine than the other jelly recipes, because there is no fruit and therefore no pectin to help set it naturally.

For the coffee jelly:

12 leaves gelatine

2 tablespoons instant coffee

4 tablespoons caster sugar

For the brown bread ice cream:

55g unsalted butter

85g wholemeal breadcrumbs

170g caster sugar

8 egg yolks

1 vanilla pod, split

570ml single cream

2 tablespoons Madeira

Soak the gelatine leaves in a bowl of ice-cold water until soft (the water must be cold or the gelatine will dissolve), then take out and squeeze.

Put the coffee and sugar in a pan with 1 litre of water, bring to the boil and stir until the coffee and sugar have dissolved. Add the gelatine and stir until dissolved. Remove from the heat and leave to cool slightly, then pour into a jelly mould or flan ring and leave in the fridge to set for at least 4 hours, or overnight.

Heat the butter in a frying pan, put in the breadcrumbs and fry until crisp. Add half of the caster sugar and let the mixture caramelise and turn golden brown. Pour into a tray in a shallow layer and leave to set, then bash with a rolling pin to break it up into fine crumbs.

In a large bowl, beat together the egg yolks and remaining sugar until thick and creamy (this is easiest with an electric hand mixer).

Put the vanilla pod and cream in a heavy-based pan and bring to the boil. Take the pan from the heat and remove the vanilla pod, then pour onto the beaten egg and sugar mixture, whisking all the time.

Pour this mixture into a heatproof bowl, set over a pan of simmering water (make sure that the base doesn't touch the water) and stir all the time until it thickens enough to coat the back of a wooden spoon. Leave to cool, then stir in the Madeira.

Churn in an ice-cream maker until starting to set, then stir in the caramelised breadcrumbs and finish. Alternatively, place in a freezer container, put into the freezer and take out and stir at 30-minute intervals until half frozen. Again, when the ice cream is starting to set, stir through the caramelised breadcrumbs and continue freezing.

Serve the ice cream with the coffee jelly.

SERVES 6

Vanilla Rice Pudding with Apple and Blackberry Compote

I know a rice pudding might be another quite old-fashioned recipe, but often the classics become so for a reason. I find the familiarity of the simple flavours restorative and comforting. Of course you can make it at any time of the year and serve it on its own or with whatever fruit you like, but it is particularly good with apples and blackberries in the autumn.

You might think that this method of making rice pudding looks difficult because it is cooked in a bowl over a pan of simmering water (a bain-marie), but in fact it is much easier than on a hob, as you barely have to stir it and you don't have to watch it to make sure it isn't catching during the slow cooking (this gentle cooking is what makes the pudding so soft and creamy). All you need to do is check now and then that the water in the pan below the bowl isn't boiling away.

150g risotto rice

75g sugar

500ml milk

10g butter

1 vanilla pod, split in half and seeds scraped out

1 teaspoon ground nutmeg

300ml double cream

For the compote:

2 small cooking apples, peeled, cored and chopped

60g sugar

juice of ½ a lemon

150g blackberries

Put the rice, sugar, milk, butter, vanilla and nutmeg into a large ovenproof bowl, cover with clingfilm and place over a pan of gently simmering water – make sure the base of the bowl isn't touching the water. Cook for about 1¾–2 hours, stirring occasionally and making sure that the water in the pan doesn't evaporate away – top it up as necessary. It is ready when most of the liquid has been absorbed, the rice is soft and the pudding has thickened. Take off the heat, stir in the double cream and remove the vanilla pod.

While the rice is cooking, make the compote: put the apples, sugar and lemon juice into a small pan with 3 tablespoons of water, cover and bring to the boil, stirring occasionally until the apples soften. Stir in the blackberries, cook for a few more minutes until the berries 'bleed', then take off the heat. Serve the rice pudding in individual bowls, with a spoonful of compote on top.

Three Sweet Tarts

Just as savoury tarts are served all year round in the farmshop cafés, our chefs also make different sweet tarts as each season's fruit arrives in the garden, usually with fruit set in frangipane – an almond cream, or scattered with a crumble topping – though when the weather gets cold, one of our favourites is the salted ginger treacle tart on page 269.

Sweet Pastry

This makes enough for one 20cm tart case, 3–4 cm deep.

150g plain flour, plus extra for rolling out

a pinch of salt

80g caster sugar

100g butter

1 whole egg, beaten

1 egg yolk, for brushing the pastry

Sift the flour into a bowl and add the salt and the sugar. Grate in the butter and mix lightly with the tips of your fingers until the mixture resembles breadcrumbs, ensuring there are no lumps of butter in the mix (alternatively you can do this using a food processor). Add the egg and mix until you have a dough – taking care not to overwork it. Form into a ball, wrap in clingfilm, and chill in the fridge for at least 30 minutes before using.

Preheat the oven to 160°C/gas 3.

Lightly flour your work surface and roll out your ball of pastry into a circle big enough to line a 20cm x 4cm deep flan tin with a removable base, leaving enough pastry to overhang the sides. Wrap the pastry carefully around your rolling pin to lift it and drop it carefully into the flan tin, pushing it gently into the base and sides of the tin – don't trim the overhanging pastry. Put the tin on a baking tray – this makes it easier to move it around – then into the fridge to rest and chill for 30 minutes (to help prevent the pastry shrinking during baking).

When ready to blind-bake, prick the base of the pastry case with a fork, line with greaseproof paper – crinkle it up first to soften it and avoid it denting the pastry – and fill with baking beans. Put into the oven for about 30 minutes, until light golden brown, then take out, remove the paper and baking beans (you no longer need these), and brush all over the inside of the pastry case with the beaten egg to seal any little holes.

Put the tin back into the oven for a further 5–10 minutes, until the base is fully baked and golden brown. Don't be scared of taking the pastry to this point. The key to a good tart base is to hold your nerve, and colour and crisp the pastry to the stage at which you would like to eat it, as once you put in your filling and return it to the oven it won't colour any more, except maybe a little around the edges, and the base will stay crispy and flaky as the filling cooks. If you only lightly colour the pastry, and the base isn't fully baked, it will be soft and doughy, making the whole tart seem heavy.

Remove from the oven and, when cool, carefully trim off the overhanging pastry with a small, sharp, serrated knife. Now you can make whichever filling you like, and bake your tart according to the instructions in each recipe.

SERVES 6

Blackberry and Apple Crumble Tart

A classic combination, best served with either custard or cream.

1 blind-baked 20cm x 4cm sweet pastry tart case (see page 266)

For the crumble:
110g flour
40g brown sugar
90g butter
55g hazelnuts, chopped
40g rolled oats

For the apple and blackberry filling:
40g butter
400g Bramley apples, peeled and cored
juice of 1 lemon
50g sugar
300g blackberries

Preheat the oven to 160°C/gas 3.

To make the crumble, put all the ingredients except the oats into a blender and blitz to a crumbly dough, then stir in the oats.

Spread the mixture over a baking sheet and put into the oven for 20 minutes, or until golden brown. Take out and, when cool enough to handle, break up the mix with your hands until it resembles breadcrumbs and keep to one side.

Meanwhile, make the filling. Put the butter into a small pan and heat gently until foaming, then add the apples, lemon juice and sugar and stir well. When the apples start to soften, add the blackberries and take the pan from the heat.

To assemble, fill the tart case with the apple and blackberry mixture, sprinkle the crumble topping all over and put into the oven for about 20 minutes, or until the topping turns golden brown.

SERVES 6

Salted Ginger Treacle Tart

The cornflakes just gives this twist on the classic recipe an interesting crunch and texture. The tart is best served with crème fraîche, which adds a little sharpness to balance out the sweetness.

1 blind-baked 20cm x 4cm sweet pastry tart case (see page 266)
1 egg, plus 1 yolk
50g double cream
1 teaspoon sea salt
60g butter
540g golden syrup
1½ teaspoons ground ginger
2 tablespoons chopped stem ginger
70g fresh brown breadcrumbs
70g cornflakes, lightly crushed

Preheat the oven to 150°C/gas 2.

In a small mixing bowl, beat the egg, yolk, cream and salt until well combined.

Put the butter into a medium pan and melt over a medium heat until golden brown and nutty. Add the golden syrup, ground and stem ginger and warm slightly, then remove from the heat and stir into the egg mixture.

Mix in the breadcrumbs and cornflakes. Spoon into the pastry case, filling it right to the rim.

Put into the oven and bake for 50 minutes, until golden brown on top – the filling should no longer wobble if you shake the tart gently. Remove from the oven and leave to cool. Serve warm.

SERVES 6

Apricot and Almond Tart

This combines apricots with a classic almond cream (frangipane), but the chefs also use the same recipe with whatever stone fruits are in the market garden, such as plums or cherries. Serve with vanilla ice cream or a bowl of whipped cream.

1 blind-baked 20cm x 4cm sweet pastry tart case (see page 266)

7–10 fresh ripe apricots

2 tablespoons apricot jam

1 egg, plus 1 egg yolk

80g ground almonds

20g plain flour

For the frangipane:

100g butter

100g caster sugar

First make the frangipane. Cream the butter and sugar together in a mixing bowl until creamy and pale in colour, then add the egg and yolk gradually (you can also do this in a mixer). When all the egg is combined, stir in the ground almonds and the flour. Put into the fridge for 1–2 hours, to firm up, then spoon into the tart base, using the back of a spoon moistened with a little water to smooth the surface.

Preheat the oven to 170°C/gas 3.

Cut the apricots in half, discarding the stones, then, depending on the size of the fruit, cut each half into two or three wedges. Starting at the outside of the tart, evenly space the apricots in a circular fashion, pushing each wedge halfway into the almond cream, and letting the other half protrude and point upwards. Repeat with more circles, working towards the centre, until all the apricots have been used up.

Place the tin on a baking tray and put into the oven for about 30 minutes, until the frangipane is golden brown, the apricots are nice and caramelised and a skewer inserted into the centre comes out clean. Remove it from the oven and leave in the tin.

Meanwhile, put the apricot jam into a small pan with 2 tablespoons of water and leave it to bubble until you have a syrup. Brush the top of the tart with the syrup. Allow to cool slightly, then remove the tart from the tin and serve warm.

SERVES 8

Blood Orange and Polenta Cake with Orange Whipped Cream

The blood oranges give this suprisingly light cake a handsome splash of colour so it's a wonderful pudding to serve in the winter at a time when there is not much in the fruit garden apart from apples and pears.

In an ideal world, the cake mixture would use an equal number of egg yolks and whites, to keep things neat, but whereas in most savoury cooking an extra dash of this or that is all part of the interpretation of a dish, with cakes it is all about chemistry, and you have to be precise: so 2 egg yolks and 3 egg whites it is.

85g butter, softened

130g caster sugar, plus 1 tablespoon

1 teaspoon vanilla extract

2 egg yolks

120g plain flour

30g fine polenta

2 teaspoons baking powder

a pinch of salt

90ml milk

3 egg whites

sprigs of mint, to garnish (optional)

For the oranges in caramel:

2 blood oranges

180g sugar

For the orange cream:

240g double cream

75g caster sugar

juice and zest of 1 orange

Preheat the oven to 180°C/gas 4. Grease a 20cm loose-bottomed cake tin and line with baking parchment.

For the oranges in caramel, remove the skin and the pith with a small sharp knife, keeping the curve of the orange. Turn each orange on its side, then cut into slices about 4mm thick. Put the sugar into a small heavy-bottomed pan with 100ml of water and heat, stirring, until the sugar has dissolved, then continue to cook without stirring until the syrup changes from a light amber to a dark caramel. As soon as it reaches this point, take the pan off the heat and carefully pour the caramel evenly into the base of the prepared tin.

Pat the orange slices dry on kitchen paper and arrange in overlapping circles on top of the caramel (this will be the top of the cake, so try to do this as neatly as possible).

Blood Orange and Polenta Cake continued

To make the cake, put the butter, 130g of sugar and vanilla extract in a bowl and cream together until light and fluffy, then mix in the egg yolks, a little at a time, until well combined. Fold in the flour, polenta, baking powder and salt, mix well, then add the milk gradually until you have a smooth batter.

Whisk the egg whites until they reach soft peaks, then add the tablespoon of caster sugar and whisk again to form stiff peaks. Gently fold a third of this meringue into the batter to begin with, to loosen it, then carefully fold in the rest as lightly as possible (you want to keep as much air in it as possible to keep the cake nice and light).

Spoon the batter over the layer of caramel and oranges in the cake tin and smooth gently.

Put into the oven and bake for about 35–40 minutes, or until a skewer inserted into the middle comes out clean (the cake will become quite dark in colour). Take out of the oven and place the tin on a cooling rack for 10 minutes.

Meanwhile, make the orange cream by whipping the double cream and sugar together until it just holds its shape and then folding in the orange zest and juice.

To turn out the cake, place a flat plate over the top of the tin, then, holding both tightly, turn them over together so that the cake tin is sitting, base side up, on the plate. Carefully remove the tin and base. Slice the cake and serve slightly warm with the orange cream. Garnish, if you like, with a sprig of mint.

The Bakery

Artisan food production is not simply a romanticised or old-fashioned notion. It is something I have always believed in and supported and encouraged as best I can. We have to find a more sustainable, less wasteful way to feed people and the slow, considered methods of producing things like bread, cheese and jams by hand is something I value enormously and truly believe should be safeguarded and valued, not solely for the quality and flavour but for their low environmental impact.

Bread is the most wasted food in the UK, with 24 million slices thrown away each day. I find that terrifying, particularly when it's also one of the simplest foods to preserve – all that's needed is to slice a loaf and freeze it and you'll always have the basis of a quick and simple meal ready to go.

A well-made loaf is a beautiful and healthy thing, but fast industrially made bread has neither the flavour nor the nutrition and for many people it is difficult to digest. What bread needs is time, not speed. When you allow your dough to go through a slow fermentation, especially sourdough, the micro-organisms go to work and by the time the bread is baked it is perfectly digestible. Typically in our bakery at Daylesford, our doughs take 24 hours from beginning to end.

On one level bread is one of the simplest and most basic of foods, involving just four ingredients: flour, yeast, salt and water – but if you make it in the artisan way, every time you bake, something different happens. Thousands and thousands of pages of books and papers have been devoted to the science, the technique, the flour, the yeast, because we are dealing with something that is alive and complex, that takes a great deal of understanding.

The weather, the humidity, the exact temperature of your water, the length of time you leave the dough to prove, the way you shape it... everything has an effect. Even in the same room, if you have three different people shaping the same dough, you will get three subtly different batches of loaves, especially when you are making a sourdough-style bread, built on a leaven (natural yeasts). An artisan baker has to go back to the old-fashioned skills and principles of bread-making, and understand, feel and control a complex population of micro-organisms which, one day, might be in good shape and another might be tired, just like human beings.

There are people who believe that you can't bake bread without strong white flour and that the gluten isn't strong enough in other wheats, but our bakery tells me that's not true. They make beautiful baguettes with plain flour. They just have to use a long fermentation and fold the dough during the proving, which strengthens the gluten, and in the end they get a great result. Plain flour has a good flavour and because it is lower in gluten, it is more digestible.

I feel extremely fortunate to have such a passionate team of bakers at Daylesford. Everyone cares so much about what they do and they are really proud of our breads. They are constantly innovating and creating new and interesting breads, incorporating cheeses from the creamery, or root vegetables, such as swedes, turnips, carrots and beetroot, or whatever Jez suggests might be good from the market garden. Like everything we do, the team likes to reflect the seasonality of working with farm ingredients. Each season they will create a new loaf for the farm shops that celebrates some of the best of that season's produce. Throughout the winter we'll have a beetroot sourdough, for example, then in September, our farm honey will just have been harvested and so they'll often use that. It lends the bread a gentle sweetness, which contrasts with the tang of the mildly sour dough, and we'll add crunch with the first wet walnuts of the season – one of my favourite autumnal treasures.

Heritage grains

Similarly, the team is keen to work with a range of interesting, different, old and heritage varieties of wheats from the UK and Europe in their choice of flours. It used to be that farmers were much more diverse in their planting and would have different varieties of wheat planted in the same field, so when the grains were milled, you would have an interesting, quite complex flour. But now, monoculture has taken over and it is rarely done any more.

One of the grains the bakery likes to use a lot is spelt as it gives such a good flavour, but it can be tricky to work with when you are not used to it because the gluten is more delicate and so the dough is more fragile and needs to be kneaded less, but strengthened by folding during a long fermentation.

The fact that spelt has a fragile gluten is mainly because it was a forgotten grain since the nineteenth century and has only been rediscovered relatively recently, so it hasn't been developed in the way of modern wheats, grown for the bread industry with high levels of strong gluten as a priority.

Sadly, though, I fear for its future, as growers pick up on its fashionability and start developing new, stronger strains.

The way the grains are milled is important in how they behave once they are made into a dough. It is possible to mill flour just as well, using cylinders in the modern way, as it is in traditional stone-ground mills, provided that you keep the temperature low and don't overheat the grains.

Often millers strip out the germs of the grain before milling because if these are crushed, their oils go into the flour, which then has a shorter shelf-life as the oils can turn it rancid after about three months. Whereas if you remove the germ, the bags of flour will last for a year. But the germ is what also gives the flour individuality. If you mill the whole grains slowly and gently at a lower temperature, you keep all the goodness and individual flavours of the grains intact.

Learning to bake bread

When I spoke to the bakers they all agreed that when you start out making bread, it is best to start with strong flour, which is why that has been specified in the recipes that follow. Once you get used to the technique of slow bread-making and really understand the process of using natural leaven and only small quantities of yeast, then you can begin experimenting with more interesting flours and perhaps introducing some plain or rye flour into the mix.

Do it, as the bakers do at the bakery, step by step, seeing what happens and adapting and developing your recipes.

The focus in these recipes is on flavour and understanding – these are everyday loaves that sustain life and feed the soul. When you make something that tastes beautiful and is based on correct technique and proper knowledge, only then can you start to be a bit fancy and have fun with your presentation.

My favourite bread is still a beautifully made, plain sourdough that is crusty and chewy. I would eat happily eat that with every meal and there are times when a slice of it toasted, with some creamy salted butter from the dairy, is better than any elaborately cooked meal. So often the simplest things are the best and where bread and butter is concerned, nothing could be truer.

BREADS

Notes on baking bread

Baking your own bread can seem like something that requires a lot of time and effort but, like other types of baking, the process can be very meditative and there is a real sense of pride in your achievement. You also have the satisfaction of knowing exactly what goes into your loaf.

Our bakery team is constantly developing new breads, using seasonal ingredients, and the range changes frequently, but the recipes in this chapter are some of those that remain constant. One of the characteristics of our bread is that the bakers use their own leaven, which they have nurtured for many years and which is the natural fermentation of wild yeasts, so many of these recipes require you to make a leaven and the process takes longer. However, there are also some simpler, quicker recipes, such as the Pumpernickel, which means you can have a homemade loaf on the table much sooner.

The team also told me that it is important to practise – and to fail sometimes because that is part of the learning process. It's likely you'll have the odd loaf go wrong, but practice will help you understand the process and then you will make really good bread. They also gave me a few tips to help ensure you get the best you can out of your bread-making:

Accuracy of ingredients is important in a way that it isn't in general cooking, so weigh everything – including water.

Don't automatically dust your work surface with flour before kneading your bread. Have some in a little bowl next to where you are working, but only use it if you really need to, and then use a very light dusting. Remember that if you use lots of extra flour you are adding this to your dough and altering the ratio of the ingredients.

The quantities of yeast are approximate because the quality varies according to the brand, so find one that you like, experiment with it, then stick with that brand, so that you know how it will behave, rather than chopping and changing – there are enough variables in bread, without adding more.

A traditional proving basket is a cheap but excellent piece of equipment to have in your kitchen, to hold loaves that are not being baked in a tin while they are proving. Made of wicker and lined with cloth, the basket allows the air to circulate, so that the dough doesn't dry out and, if you dust it with flour, when you turn out the dough ready to bake, it will keep this fine dusting on the top, which will give the finished bread an attractive look.

Before you put your dough into the oven, you need to score the top with the blade of a sharp, serrated knife, quickly and cleanly, so you don't drag the dough. The cuts don't need to be deep – only about 2mm – but they will create weak points in the surface of the forming crust, which will allow the gases that build up as the dough bakes to push against the crust and expand, so the bread can rise to its full potential. If you don't make any cuts, think of it as a volcano with a lid on it. What will happen is that eventually the crust will crack randomly under the pressure, but still not enough to allow the bread to rise as much as it would like.

When you score your bread, the simplest way to do it, for round loaves, is to make a wide cross on the top, and for longer loaves, to make a series of slashes along the top of the loaf. Always make your cuts as long as possible: the full width of the bread.

Throw a handful of ice cubes into the bottom of the oven as you put in the bread. The cubes evaporate instantly, without marking your oven, and the steam they create softens the top of the dough at the beginning of baking, allowing it to rise more and give you a nice shiny crust at the end.

Squash, Honey and Sage Bread

MAKES 1 LOAF

This bread is baked at quite a low temperature because at a higher heat the squash, honey and malt extract would caramelise and colour too quickly, so you would run the risk of the bread burning and tasting bitter.

- 180g strong white flour
- 20g butter, softened
- 10g sugar
- 35g milk
- 5g fresh yeast or 3g dried yeast
- 5g fine sea salt
- 50g butternut squash, coarsely grated
- 10g pumpkin seeds
- 5g malt extract
- 1 teaspoon honey
- 4 sage leaves, very thinly sliced
- a little vegetable oil, for greasing the bowl and the clingfilm

Put the flour, butter, sugar, milk, yeast and salt into a bowl with 70g of lukewarm water and mix into a dough.

Turn out the dough on to your work surface and knead for 5 minutes.

Add the squash, pumpkin seeds, malt extract, honey and sage and knead for another 5 minutes (using extra flour if necessary), until you have a smooth and shiny dough.

Transfer to a lightly oiled bowl, cover with oiled clingfilm and allow to rise in a warm place, until doubled in size.

Turn out the dough and 'de-gas' by pressing down briefly and gently with the flat of your hand, to even out the bubbles of air, then shape into a ball. Transfer to a floured proving basket or a baking sheet. Cover with a clean tea towel and again leave in a warm place to prove, until doubled in size.

Meanwhile, preheat the oven to 180°C/gas 4. This might seem like a low temperature, however, this is a bread that colours very quickly.

If the dough has been resting in a proving basket, turn it out on to a baking sheet. If it has been resting on a baking sheet, finely dust the top with flour (if it has been in a proving basket it will already have this fine dusting).

Score the top of the loaf swiftly and cleanly with the blade of a sharp serrated knife (see page 283), put it into the oven and bake for about 25–30 minutes. If you throw some ice cubes into the bottom of the oven, this will create steam and enhance the crust.

The bread is ready when, if you tap the base, it sounds hollow.

Remove from the oven and place on a cooling rack.

Pumpernickel

MAKES 1 X 15CM X 9CM LOAF

Our bakery team advises that it will be easier to do the mixing and kneading for this loaf in a mixer with a dough hook as it is a very sticky dough. You mix the ingredients together on the first speed for 10 minutes, then knead for 3 minutes on the second speed, add the raisins and mix for another 3 minutes back on the first speed.

The bread is only proved once and the finished loaf will look more like a brick than a risen bread. This is baked at quite a low temperature, for quite a long time, as it doesn't form a crust like many of the breads.

60g cracked rye

25g pumpkin seeds, plus 1 tablespoon for topping

15g sunflower seeds, plus 1 tablespoon for topping

10g linseeds

10g fine sea salt

60g strong white flour

120g dark rye flour

1 teaspoon instant coffee

60g molasses

12g fresh yeast

75g raisins

a little vegetable oil, for greasing your hands and the clingfilm

Put the cracked rye, all the seeds (except those for the topping) and the salt into a bowl and stir in 90g of lukewarm water. Leave to soak overnight.

Have ready a 15cm x 9cm loaf tin (if it is not non-stick, grease and line with greaseproof paper).

Put both flours into a bowl with the coffee, molasses and yeast, and add the soaked seeds. Mix together, then turn out on to a lightly floured surface and knead for 10 minutes. Because the dough is naturally sticky, this is one occasion when you probably will have to use a little extra flour, as necessary. Alternatively, if you find it is really difficult to knead with your hands, leave the dough in the bowl and use a wooden spoon to move it around in a kneading action.

Sprinkle the raisins over the dough and knead again for a few minutes, again with your hands or a wooden spoon, until they are well distributed.

Pumpernickel continued

The dough will still be quite soft and sticky, so oil your hands to make it easier to handle, then form it into a rough ball, drop it into the prepared tin and press lightly down on the top. Scatter over the extra pumpkin and sunflower seeds.

Cover with a clean tea towel and leave in a warm place to prove for 1–2 hours, until it is about one and a half times the original size.

Meanwhile, preheat the oven to 160°C/gas 3. Put in the tin and bake the bread for 45–50 minutes. For this bread, check as you would for a cake. Insert a metal skewer into the centre and if it comes out clean, the bread is ready.

Turn out and cool on a rack.

MAKES 3 SMALL LOAVES OR 12 BUNS

Hot Cross Loaf or Buns

Easter is an important time of year for me – both at home and at Daylesford where we do a lot over the Easter weekend to celebrate. There are traditional family activities to get involved in on Easter Sunday, such as egg decorating and the chance to make Easter biscuits. And at home my husband and I will always put on an Easter egg hunt for our grandchildren. We'll always have lots of hot cross buns and I think Daylesford's traditional recipe is wonderful. It is great fun to make with children who can help to shape the buns and pipe the white crosses. You just need to make sure you leave enough time to prove the dough, which is essential for a light, fluffy bun. The loaf is lovely, sliced and spread with a little good butter and jam.

500g strong white flour
10g fine sea salt
10g fresh yeast or 4g dried yeast
75g butter, softened
25g golden syrup
110g milk
1 teaspoon mixed spice
2 teaspoons ground cinnamon
90g currants
70g sultanas
70g mixed peel
1 egg, beaten, for brushing

a little vegetable oil, for greasing the bowl and the clingfilm

For the crossing paste:
40g plain flour
10g sunflower oil

For the glaze:
25g sugar
1 teaspoon liquid malt (available from health stores)
1 teaspoon lemon juice

Line a large baking tray with baking parchment.

To make the dough, sift the flour and salt into a large bowl with the yeast, butter, syrup, milk and 200–220g of lukewarm water. Mix together to form a dough. Turn out on to your work surface and knead for 10 minutes, or until smooth.

Roll the dough out with a rolling pin to about 2cm thick, then sprinkle over the spices, fruit and peel. Knead again until evenly distributed throughout the dough, then form into a ball and put into a large oiled bowl. Cover with oiled clingfilm and leave to rise in a warm place for about 1–2 hours, until doubled in size.

Hot Cross Loaf or Buns continued

Turn out the dough and 'de-gas' by pressing down briefly and gently with the flat of your hand, to even out the bubbles of air, then either shape into a ball and press gently into three 400g loaf tins, or, for buns, divide into 12 pieces (for evenly sized rolls, weigh the whole dough, divide by 12 then weigh each piece to that weight). Form each one into a bun shape and place on the prepared baking tray. Cover with a clean tea towel.

Leave the loaf tins or buns in a warm place, again until doubled in size, then brush with beaten egg.

Preheat the oven to 210°C/gas 7 for buns, or 200°C/gas 6 for loaves.

While the loaves or buns are proving, make the crossing paste by mixing the flour and oil with 40g water. Put it into a piping bag with a medium, plain nozzle.

To make the glaze, put the sugar, malt and lemon juice into a pan with 25g water and heat, stirring, until boiling, then take off the heat and set aside to cool.

When the loaves or buns have doubled in size, pipe 'crosses' on the top of each one (one big cross for the loaves), then put into the oven and bake for about 20–30 minutes, until golden. If you throw some ice cubes into the bottom of the oven, this will create steam and enhance the rising and look of the buns or loaves.

Remove from the oven (turn the loaves out of their tins) and, while still warm, brush the tops with glaze. Leave to cool on a rack or racks.

Baking with Natural Leaven (a Starter)

The oldest breads were fermented naturally using wild yeasts, long before the first commercial versions became available in the nineteenth century. Every baker has their own preferred way of making a leaven (sometimes also called a starter, mother or ferment), which is a mixture of flour and water, and an ingredient like honey, yoghurt or fruit, in this case, grapes, which feeds the wild yeasts and boosts the fermentation. When you get used to working with these natural yeasts, you can evolve your own leaven in the way you want. These things take time and patience, though. Once you have made your leaven, you will need to refresh or 'feed' it with more flour and water every 2 days and it will continue to grow, so unless you bake regularly – using some of the leaven each time – it will start to take over your fridge like a live monster. The answer? Keep baking.

Here is how to make your own leaven. The process will take around 7 days from start to finish.

Days 1–4
Start with 800g of grapes. Remove the stalks, then blend until smooth, put into a bowl and cover with clingfilm.

Leave in a warmish room at around 24–30°C for 3 days, during which time it will start to ferment. By day 3 it should be bubbling, and have an alcoholic, acidic smell.

Strain through a very fine sieve or coffee filter into a jug.

Days 4–5
Measure 200g of the fermented grape juice and put into a bowl. Mix in 200g of strong flour until smooth.

Cover with clingfilm and leave for 24 hours in a warm room (24–30°C again), after which the mixture should have fermented, have lots of bubbles and have grown in size.

Days 5–6
Take 2 large tablespoons of the fermented mixture and weigh it. Add enough lukewarm water to bring the total up to 200g and whisk until smooth. Add 200g of strong flour and mix until well blended.

Cover with clingfilm and leave in a warm room (24–30°C) for another 24 hours.

Days 6–7
Take 2 large tablespoons of the fermented mixture and weigh it. But this time, you need to lower the hydration of the leaven, ready for bread-making. So add only enough lukewarm water to bring the total up to 120g and whisk until smooth. Then add 200g of strong flour and mix until well blended.

Cover with clingfilm and leave in a warm room (24–30°C) for 1–2 hours, then put into the fridge for 24 hours to stabilise it and finish the maturation as it cools down. By the next day it will be ready to use.

Day 7
You will now have 320g of your own leaven. From now on, you can use some for baking and what is left over can be kept going indefinitely in a bowl in the fridge, as long as you continue to feed it every 2 days, in the same final ratio of 100 per cent flour and 60 per cent liquid. If you don't keep refreshing and re-activating it in this way, it will lose its properties.

All the following recipes are based on this natural leaven.

Sourdough

MAKES 1 LOAF

When you make bread with natural leaven, it has an element of sourness to it that gives this style of bread its name. However, the degree of sourness and the style of bread varies from bakery to bakery. There are two kinds of acidity in a natural leaven, acetic and lactic. The acetic is the one you find in vinegar and is quite sharp, whereas the lactic is mainly found in yoghurt and is more subtle, bringing a sensation of freshness to the bread without too much sourness, so it really stimulates the tastebuds.

- 530g strong white flour
- 10g fine sea salt
- 200g natural leaven (see page 294)
- a little vegetable oil, for greasing the bowl and the clingfilm

Put the flour into a bowl with the salt and 350g of lukewarm water and mix to a dough.

Turn out the dough on to your work surface and knead for 5 minutes. Add the leaven and knead for another 5 minutes (using extra flour if necessary).

Transfer to a lightly oiled bowl, cover with oiled clingfilm and allow to rise in a warm place until it is one and a half times its original size.

Turn out the dough and 'de-gas' by pressing down briefly and gently with the flat of your hand to even out the bubbles of air, then shape into a ball. Transfer to a floured proving basket or a baking sheet. Cover with a clean tea towel and again leave in a warm place to prove, until it has increased by one and a half times in size.

Preheat the oven to 240°C/gas 9.

If the dough has been resting in a proving basket, turn it out on to a baking sheet. If it has been resting on a baking sheet, finely dust the top with flour (if it has been in a proving basket it will already have this fine dusting).

Score the top of the loaf swiftly and cleanly with the blade of a sharp serrated knife (see page 283), then put it into the oven and bake for about 30 minutes. If you throw some ice cubes into the bottom of the oven when the bread goes in, this will create steam and enhance the crust.

The bread is ready when, if you tap the base, it sounds hollow.

Remove from the oven and place on a cooling rack.

Seven Seeds Sourdough

MAKES 1 LOAF

The malted grain in the flour that our bakers use is the seventh seed.

120g strong white flour	10g millet seeds
120g malted flour	10g sunflower seeds
10g pumpkin seeds	10g fine sea salt
10g sesame seeds	100g natural leaven (see page 294)
10g linseeds	a little vegetable oil, for greasing the bowl and clingfilm
10g poppy seeds	

Put all the ingredients, apart from the leaven and the oil, into a bowl with 175g of lukewarm water and mix to a dough.

Turn out the dough on to your work surface and knead for 5 minutes. Add the leaven and knead for another 5 minutes (using extra flour if necessary). Transfer to a lightly oiled bowl, cover with oiled clingfilm and allow to rise in a warm place until one and a half times its original size.

Turn out the dough and 'de-gas' by pressing down briefly and gently with the flat of your hand to even out the bubbles of air, then shape into a tight bloomer (oval shape) and transfer to an oval-shaped floured proving basket or a baking sheet. Cover with a clean tea towel and again leave in a warm place to prove, until it has increased by one and a half times in size.

Preheat the oven to 240°C/gas 9.

If the dough has been in a proving basket, turn it out on to a baking sheet. If it has been on a baking sheet, finely dust the top with flour (if it has been in the proving basket, it will already have a fine dusting).

Score the top of the loaf swiftly and cleanly with the blade of a sharp serrated knife (see page 283), then put into the oven and bake for about 35 minutes. If you throw some ice cubes into the bottom of the oven when the bread goes in, this will create steam and enhance the crust.

The bread is ready when, if you tap the base, it sounds hollow. Remove from the oven and place on a cooling rack.

Two Flowerpot Breads

These are fun to make and look fantastic. Our bakery team bakes them at quite a low temperature because a flowerpot is not designed for baking and doesn't transmit the heat that well, so at a high temperature the top of the loaf always has the potential to brown and become over-baked before the rest is ready.

Red Onion, Cheddar and Chilli Bread

The almonds will help to protect the top of the bread, but if you find that it becomes a little too brown the first time you try this, next time dust the top of the dough with a little flour before baking.

Go lightly on the chilli and paprika, as these are meant to bring a delicate flavour and not dominate the other ingredients or add any real heat, in the case of the chilli.

In this bread we use less leaven with a small quantity of fresh yeast, to help the lightness.

180g strong white flour

45g dark rye flour

5g salt

2g fresh yeast

90g natural leaven (see page 294)

35g chopped red onion

35g mature Cheddar cheese, grated

a good pinch of paprika

a good pinch of chilli flakes

15 flaked almonds

a little vegetable oil, for greasing the bowl and clingfilm

Put the flours, salt and yeast into a bowl with 160g of lukewarm water and mix to a dough.

Turn out the dough and knead for 5 minutes. Add the leaven and all the rest of the ingredients, except for the almonds and the oil, and knead for another 5 minutes.

Transfer to a lightly oiled bowl, cover with oiled clingfilm and allow to rise in a warm place until one and a half times its original size.

Have ready a clean terracotta flowerpot, lined with greaseproof paper.

Turn out the dough and 'de-gas' by pressing down briefly and gently with the flat of your hand to even out the bubbles of air, then shape into a ball and press gently into the lined flowerpot. Cover with a clean tea towel and again leave in a warm place to prove until it has doubled in size.

Preheat the oven to 180°C/gas 4.

Brush the top of the dough with a little water and sprinkle on the almonds, then put into the oven and bake for about 35–40 minutes.

If you throw some ice cubes into the bottom of the oven when the bread goes in, this will create steam and enhance the crust.

Tapping the base of this bread to check that it sounds hollow isn't such a good guide for this as with other breads because the pot will help to keep in more moisture, but after 35–40 minutes it will be baked.

Remove from the oven, turn out of the flowerpot and place on a cooling rack.

Nettle Bread

Nettles are a perfect example of sustainable, nutritious food, full of proteins, fibres and minerals, and have been valued since antiquity, but make sure you harvest them from your own garden or from a clean place, free of pesticides, and not from the side of the road, where they can be polluted. You need to wash and dry them, as you would salad leaves (it is best to wear gloves, as they sting), then lay them on a tray in a very cool oven (at its lowest setting) for a few hours until they are perfectly dry. Then it will be easy to shred the leaves.

260g strong white flour

6g salt

100g natural leaven (see page 294)

6g washed and completely dried nettle leaves, shredded (see introduction, above)

a pinch of anise seeds

a little vegetable oil, for greasing the bowl and clingfilm

Put the flour and salt into a bowl with 180g of lukewarm water and mix to a dough.

Turn out the dough and knead for 5 minutes. Add the leaven, nettles and seeds, and knead for another 5 minutes.

Transfer to a lightly oiled bowl, cover with oiled clingfilm and allow to rise in a warm place until one and a half times its original size.

Have ready a clean terracotta flowerpot, lined with greaseproof paper.

Turn out the dough and 'de-gas' by pressing down briefly and gently with the flat of your hand to even out the bubbles of air, then shape into a ball and press gently into the lined flowerpot. Cover with a clean tea towel and again leave in a warm place to prove until it has increased by one and a half times in size.

Preheat the oven to 180°C/gas 4.

Finely dust the top of the dough with flour, then put into the oven and bake for about 35–40 minutes.

If you throw some ice cubes into the bottom of the oven when the bread goes in, this will create steam and enhance the crust.

Tapping the base of this bread to check that it sounds hollow isn't such a good guide for this as with other breads because the pot will help to keep in more moisture, but after 35–40 minutes it will be baked.

Remove from the oven, turn out of the flowerpot and place on a cooling rack.

Notes on cakes and breaks

English cakes have stood the test of time for a reason: everyone loves the classics. How often do you hear people say that their favourite is a lemon drizzle or a coffee and walnut cake?

At Daylesford we love the classics and we rarely deviate too far from traditional recipes but the bakers will sometimes give traditional recipes a contemporary edge and a lightness of touch, adding less sugar or creating a softer, more airy sponge.

Cake time is also a good opportunity to make a refreshing tea, such as fresh mint – just leaves and hot water – or the Daylesford favourite, lemon and ginger, which we make with two slices of fresh lemon, two teaspoons of freshly grated ginger (with the skin left on), mixed into a cup of hot water, with a tablespoon of honey served separately on the side.

Our brownies have become quite legendary and are now a staple in our farmshops. And biscuits are always real crowd-pleasers, made to tried-and-tested recipes that we have been using for many, many years.

Chocolate Cake

MAKES 1 X 23CM OR 25CM ROUND CAKE

The bakers used to make a quite grown-up and very bitter chocolate cake, which was very popular with adults, but not so much with children, so this is a slightly sweeter one that everyone likes.

460g plain flour

a pinch of salt

1 teaspoon bicarbonate of soda

1 teaspoon vanilla extract

420ml milk

280g good dark chocolate (at least 70% cocoa solids), broken up into squares

380g butter, softened

330g caster sugar

330g light soft brown sugar

5 large eggs

For the filling/glaze:

300g good dark chocolate (at least 70% cocoa solids), broken up into squares

125g butter, roughly chopped

75g golden syrup

4 tablespoons sunflower oil

Preheat the oven to 180°C/gas 4 and grease and line two 23 or 25cm deep springform cake tins with baking parchment.

Combine the flour, salt and bicarbonate of soda. Stir the vanilla extract into the milk.

Put the chocolate into a heatproof bowl and place over a pan of gently simmering water (make sure the base of the bowl doesn't touch the water). Stir and, when melted, remove from the heat.

In a very large mixing bowl, cream the butter, caster sugar and brown sugar together using an electric hand whisk or in a freestanding mixer until pale and fluffy and then gradually beat in the eggs. Gently stir in the melted chocolate. Fold in half the flour, then half the milk, then fold in the rest of the flour and finish with the rest of the milk, making sure that all the flour has been fully incorporated into the mixture.

Spoon into the prepared tins and bake in the oven for 1 hour, or until a skewer inserted in the middle of the cakes comes out clean. Transfer the tins to a cooling rack and leave to cool for 20 minutes. Remove the springform tin rings, but leave the cakes on the tin bases, sitting on the wire rack, to cook completely.

Chocolate Cake continued

Meanwhile, to make the filling/glaze, put the chocolate into a heatproof bowl with the butter and golden syrup, place over a pan of gently simmering water (again, make sure the base of the bowl doesn't touch the water) and leave to melt. Once the chocolate and butter have melted, stir to combine everything together until smooth.

Take off the heat, then spoon off about a third of the mixture into a separate bowl. Pour the sunflower oil into the larger quantity and stir in. Allow both mixtures to cool.

Use a spoon to lightly whisk the smaller quantity of the chocolate mixture until it thickens slightly.

Remove the metal bases and baking parchment from both cakes and sit them directly on the cooling rack. Spread the thickened filling over the top of one, then put the other cake half on top to sandwich them together.

Check that the remaining glaze is still pourable and, if not, put the bowl back over a pan of gently simmering water, as before, for just long enough to loosen it.

Keep the sandwiched cake on the rack, but put a plate or tray underneath to catch the drips, then carefully pour and spread the remaining glaze over the top of the cake, using a palette knife to ensure it also coats the sides. If some of the glaze falls onto the plate, scoop it up and spread it over the sides and into the middle between the cakes, so they are completely covered. Very carefully transfer to a large serving plate and serve.

Manuka Honey Cake

MAKES 1 X 18CM ROUND CAKE

All good honey is said to have restorative, even healing properties, but certain batches of Manuka honey from New Zealand have been shown to have special antibacterial qualities – the Maoris traditionally put it on cuts to help fend off infection. This cake has a very distinctive, acquired flavour, quite potent and floral, from the Manuka; other honeys would just tend to make this cake sweet without adding notable flavour.

160g butter, plus a little extra for greasing the tin
100ml Manuka honey
100ml clear honey
65g light brown sugar
2 large eggs, beaten

220g self-raising flour

For the syrup:
2 tablespoons Manuka honey
1 tablespoon sugar

Preheat the oven to 170°C/gas 3. Grease and line a round 18cm cake tin that is about 6cm deep.

Melt the butter in a pan, then take off the heat, add the honeys and sugar, and stir until the sugar has dissolved.

Beat the eggs in a bowl, pour in the honey mixture, mix well, then whisk in the flour briefly until just combined. Pour into the prepared cake tin.

Put into the oven and bake for about 40–50 minutes, or until a skewer inserted into the centre comes out clean. Remove from the oven and leave in the tin on a rack to cool slightly before turning out. Once turned out, return the cake to the rack, but put a plate underneath to catch any drips when you glaze it with syrup.

Make the syrup by putting the honey and sugar into a pan with 2 tablespoons of water and bringing to the boil. Take off the heat straightaway and pour over the cake while it is still warm.

MAKES 1 X 1KG LOAF CAKE

Spiced Apple Cake with Streusel Topping

Streusel is a German name for a crumble-style topping made with flour, sugar, butter and, in this case, chopped nuts, which is scattered over pies and cakes, particularly apple – like this quite autumnal, spiced apple cake. As the cake bakes, the juice will come out of the apples and make the mixture quite moist, so be sure to do the skewer test to check that it is baked through, otherwise, the cake will be doughy in the centre.

a little butter, for greasing the tin
2 eggs
100g caster sugar
120ml sunflower oil
120ml milk
150g plain flour
40g wholemeal flour
1 teaspoon bicarbonate of soda
2 teaspoons baking powder
2 teaspoons ground cinnamon
1 teaspoon ground nutmeg
160g peeled, cored and chopped apple (approximately 2 apples)
70g sultanas

For the streusel topping:
60g plain flour
60g light brown sugar
40g butter
40g roughly chopped pecan nuts

Preheat the oven to 160°C/gas 3. Grease a 1kg loaf tin with butter.

To make the topping, put the flour, sugar and butter into a bowl, rub together until the mix resembles breadcrumbs, then stir in the pecan nuts. Put into the fridge to chill and harden for 30 minutes.

Meanwhile, to make the cake batter, whisk together the eggs and sugar in a bowl until combined but not aerated, then slowly add the sunflower oil and milk, whisking until they are mixed in. Finally fold in the flours, bicarbonate of soda, baking powder, cinnamon and nutmeg, until you have a smooth batter. Fold in the apples and sultanas and pour into the prepared loaf tin.

Take the streusel topping from the fridge and scatter over the top. Put into the oven and bake for about 25–30 minutes, until a skewer inserted into the centre comes out clean. Take out of the oven and leave to cool in the tin before turning out.

Earl Grey Cake

MAKES 1 X 18–20CM ROUND CAKE

In this recipe, Earl Grey tea is used instead of alcohol or juice to soak and plump up the fruit, so that the cake is kept moist. The tea has a strong aroma initially, but tones right down and just gives gentle floral background notes to the quite light cake. For the best flavour, use loose-leaf tea.

3g Earl Grey tea or 3 Earl Grey tea bags

200g raisins

200g sultanas

200g currants

200g plain flour

1 teaspoon baking powder

175g butter, softened

175g light brown sugar

3 eggs, beaten

1 teaspoon vanilla extract

Brew the Earl Grey tea by placing the tea leaves or tea bags in a large jug and pouring over 500ml of boiling water. Leave for 5 minutes. Put the dried fruit into a bowl, then strain the tea through a fine sieve over the top (or just remove the tea bags, if using). Cover (with muslin, ideally) and leave overnight, stirring occasionally when you can, to ensure the fruit absorbs the liquid.

Preheat the oven to 170°C/gas 3. Grease and line an 18–20cm round cake tin and sift together the flour and baking powder.

In a bowl, cream the butter and sugar until pale and fluffy. Gradually beat in the eggs and vanilla extract. If the mix looks like it's 'curdling', add a teaspoon of the measured flour.

Drain the soaked fruit and stir into the mixture. Gently fold in the flour and baking powder, then spoon into the prepared cake tin and smooth the top.

Bake in the oven for about 1½ hours, or until a skewer inserted into the cake comes out clean. Leave the cake to cool on a rack before removing the tin.

Lemon Drizzle Cake

MAKES 1 X 15CM X 9CM LOAF CAKE

We have been making this since the farmshop opened. It is a true English classic, but there are many different opinions on how to make it. Often it is done with a glaze of lemon icing over the top, whereas in a true drizzle cake, the lemony sugary syrup should seep into the cake itself. The secret is to pour or 'drizzle' the warm syrup over the cake while it is still warm, so that the cake absorbs the flavour. Eat within a week, otherwise you lose that zingy lemony freshness.

2 large eggs
130g caster sugar
80ml double cream
50g butter, melted and cooled
125g plain flour
1 teaspoon baking powder
zest of 1 lemon

For the syrup:
juice of 1 lemon
80g caster sugar

Preheat the oven to 170°C/gas 3. Grease and line a 15cm x 9cm loaf tin.

In a bowl, whisk together the eggs and sugar until the sugar has just dissolved. Stir in the cream and the melted and cooled butter, mix well, then add the flour, baking powder and lemon zest and whisk briefly until well combined.

Spoon into the prepared tin and bake in the oven for 30–40 minutes, until the cake is golden brown, well risen, firm to the touch and a skewer inserted into the centre comes out clean. Remove from the oven and leave the cake, still in its tin, on a cooling rack.

Meanwhile, make the syrup by combining the lemon juice and sugar in a small pan and stirring over a low heat until the sugar has just dissolved. Remove from the heat. Prick the top of the cake all over with a skewer and pour the syrup over it while still warm. Leave until the cake has cooled, then remove from the tin.

MAKES ABOUT 16

Dark and White Chocolate Brownies

These brownies have become something of a Daylesford icon and they're a staple at the bakery and in our farmshops. A crisp, sugary topping gives way to the soft, smooth dark chocolate filling which is contrasted with a lovely crunch from the shards of white chocolate that run through the middle. Decadent and rich, these are very hard to resist.

4 eggs

150g caster sugar

150g demerara sugar

350g unsalted butter

350g good dark chocolate (at least 70% cocoa solids)

180g plain flour

1 teaspoon baking powder

1 teaspoon vanilla extract

380g white chocolate buttons

Preheat the oven to 160°C/gas 3 and line a 30cm square cake tin (or equivalent) with baking paper.

In a bowl, whisk together the eggs and both sugars until thick.

Put the butter and dark chocolate into a heatproof bowl over a pan of simmering water (make sure the base doesn't touch the water) and let the chocolate melt.

Let the chocolate cool a little, then add to the beaten eggs and sugar, and mix well. Gently fold in the flour, baking powder and vanilla extract and, when thoroughly combined, add the chocolate buttons and mix in gently.

Spoon into the prepared tin and bake in the oven for 20 minutes, making sure you don't overbake the brownies. They need to stay quite soft and moist in the middle, so they should be springy to the touch and, if you insert a skewer into the centre, it should come out sticky, not clean, as you would expect with a cake.

Take out of the oven and allow to cool in the tin before turning out on to a board or clean work surface and cutting into squares.

Energy Bars

MAKES 12 BARS

We wanted to create something nice, light and healthy that visitors to the farmshop could have with a coffee or tea – and this was the result, a chewy, seedy, fruity bar.

160g porridge oats	90g honey
3 tablespoons sesame seeds	100g golden syrup
3 tablespoons sunflower seeds	100g butter, softened
2 tablespoons pumpkin seeds	160g light brown sugar
2 tablespoons flax seeds	100g sultanas
2 tablespoons poppy seeds	80g dried apricots, chopped
70g desiccated coconut	1 teaspoon bicarbonate of soda
70g light rye flour	

Preheat the oven to 170°C/gas 3. Grease and line a 22cm square tin that is about 5 cm deep.

Put the oats, 2 tablespoons of sesame seeds, 2 tablespoons of sunflower seeds, the pumpkin, flax and poppy seeds, the coconut and the rye flour into a mixing bowl.

Put the honey, syrup, butter and sugar into a small pan and heat gently. Add the sultanas and apricots and continue to heat until just bubbling, then take off the heat, stir in the bicarbonate of soda and pour over the oat mixture. Mix well, pour into the prepared tin, spread evenly and then sprinkle with the remaining sesame and sunflower seeds.

Bake in the oven for 35 minutes, then turn off the oven, open the door and leave the energy bars inside to dry out for 10 minutes. Remove the tin from the oven. Allow to cool before cutting into bars.

Three Cheese Biscuits

These little savoury biscuits are good to nibble on any time or to put out with drinks.

Cheddar Biscuits

250g strong white flour, plus extra for rolling out

½ teaspoon baking powder

½ teaspoon sea salt

150g cold butter, grated

2 tablespoons double cream

1 egg yolk

150g Cheddar cheese, grated

Put the flour, baking powder and salt into a bowl with the butter, and rub together with your fingertips until the mixture resembles breadcrumbs. Add the cream, egg yolk and 100g of the cheese, and bring together gently into a dough.

Turn out on to a lightly floured work surface and divide the dough in half. Roll each half into a log shape, about 4cm in diameter. Wrap each one in clingfilm and put into the fridge to chill and firm up for about 30 minutes.

Preheat the oven to 180°C/gas 4.

Take the logs of dough from the fridge and slice each one into 8 rounds. Lay them on a non-stick baking sheet (or alternatively line one with greaseproof paper) and put back into the fridge to rest again for another 30 minutes. Bake in the oven for about 15 minutes, until golden brown.

Take out of the oven and, while still hot, scatter the tops with the remaining cheese. Leave to cool on the tray for a few minutes before transferring to a cooling rack.

Blue Cheese and Walnut Biscuits

325g strong white flour, plus extra for rolling out

1 teaspoon baking powder

½ teaspoon salt

190g cold butter, grated

80ml double cream

250g blue cheese, grated (or chopped, if soft)

1 large egg yolk

125g walnuts, chopped

Put the flour, baking powder and salt into a bowl with the butter, and rub together with your fingertips until the mixture resembles breadcrumbs. Mix in the cream, blue cheese and egg yolk to form a dough.

Turn out on to a lightly floured work surface and divide the dough in half. Roll each half into a log shape, about 4cm in diameter. Wrap each one in clingfilm and put into the fridge to chill and firm up for about 30 minutes.

Preheat the oven to 180°C/gas 4.

Take the logs of dough from the fridge and slice each one into 12 rounds. Lay them on a non-stick baking sheet (or alternatively line with greaseproof paper) and top each round with some walnuts, pressing the nuts down slightly. Put the baking sheet back into the fridge to chill again for 30 minutes. Bake in the oven for about 15–20 minutes, until golden brown.

Leave to cool on the tray for a few minutes before transferring to a cooling rack.

Parmesan, Chilli and Marcona Almond Biscuits

170g butter, softened

100g Parmesan cheese, finely grated

75g mature Cheddar cheese, finely grated

170g plain flour, plus extra for rolling out

½ teaspoon sea salt

½ teaspoon smoked paprika

¼ teaspoon dried chilli flakes

150g roasted, salted Marcona almonds, roughly chopped

Put the butter, Parmesan and Cheddar into a bowl and cream together. Mix in the flour to form a stiff dough. Add the salt, paprika, chilli flakes and almonds and, when incorporated into the dough, turn out on to a lightly floured work surface. Divide the dough into 6 pieces and roll out each into a log shape, about 4cm in diameter.

Wrap each log in clingfilm and put into the fridge to chill for at least 30 minutes.

Preheat the oven to 180°C/gas 4.

Take the logs of dough from the fridge and slice each log into rounds 0.5cm thick. Lay the rounds on a non-stick baking sheet or sheets (or alternatively line with greaseproof paper) and put back into the fridge to rest again for another 30 minutes. Bake in the oven for 10–15 minutes, until golden brown and crunchy.

Leave to cool on the tray for a few minutes before transferring to a cooling rack.

Lemon Shortbread Biscuits

MAKES 10–12 BISCUITS

Sometimes we dip the finished biscuits into melted white chocolate, which makes them very decadent. Don't overwork the dough, as shortbread should be crumbly and buttery, almost melt-in-the-mouth. The more usual way to make shortbread is to pack the mixture into a baking tin, then cut it into fingers once it is baked. However, because we like to make round (or sometimes heart-shaped) biscuits, in advance of baking we put the dough into the fridge to firm up before stamping it out with a biscuit cutter – otherwise the dough would just fall apart.

125g butter, softened

65g caster sugar, plus a little extra for sprinkling on top

115g plain flour, plus extra for rolling out

65g rice flour

a pinch of salt

zest of 1 lemon

In a bowl, cream the butter with the sugar until pale and fluffy. Add the flours, salt and lemon zest and combine until the mixture comes together. Form this dough into a ball. Lightly flour a work surface and roll out the dough until it is 0.5cm thick. Lift carefully on to a flat baking sheet and put into the fridge for 1 hour to firm up.

Preheat the oven to 170°C/gas 3 and line a baking tray with greaseproof paper.

Take the dough from the fridge, slide it off the baking sheet on to your work surface, then, using a biscuit cutter of about 7cm in diameter, cut out 10–12 rounds (or similar-sized shapes of your choice). Lay the rounds on the prepared baking tray.

Bake in the oven for 10–15 minutes, until a very light golden brown. Take out of the oven and leave the biscuits on the tray for a few minutes to firm up before carefully lifting them off with a fish slice or spatula and transferring them to a wire rack to cool. Dust with caster sugar before serving.

Ginger Biscuits

MAKES ABOUT 20 BISCUITS

These contain stem ginger, so although they are crunchy on the outside they have little pockets of soft, sticky gingeriness on the inside.

- 150g butter, softened
- 100g caster sugar
- 3 tablespoons dark brown sugar
- 2 tablespoons molasses
- 2 whole eggs
- 170g plain flour
- 2 teaspoons bicarbonate of soda
- 2 teaspoons ground ginger
- 1 teaspoon ground cinnamon
- 1 teaspoon mixed spice
- 2 tablespoons stem ginger, chopped small (about 0.5cm)
- a pinch of salt
- caster sugar, for dusting

Preheat the oven to 180°C/gas 4 and grease two baking trays with butter.

Put the butter and sugars into a bowl and cream together until pale and fluffy. Add the molasses, then gradually beat in the eggs.

Fold in the flour, bicarbonate of soda, spices, stem ginger and salt.

Spoon dessertspoonfuls of the mixture on to the prepared baking trays and flatten slightly into rounds of about 1cm thick with the back of the spoon.

Put into the oven and bake for about 10–15 minutes, until golden. Remove from the oven and lightly sprinkle with caster sugar while still warm.

Allow to cool slightly on the trays before removing to a cooling rack.

Milk Chocolate, Almond and Espresso Fudge

This recipe makes a smooth, simple fudge, which is not cloying. Coffee and chocolate are natural partners – but the coffee element is quite subtle. If you don't have an espresso maker, make a very strong filter coffee using 5 parts boiling water to 1 part coffee.

400g good milk chocolate

500ml double cream

500g caster sugar

50g golden syrup

75g unsalted butter

75g whole blanched almonds, lightly toasted in a dry pan and roughly chopped

75g toasted flaked almonds

75ml espresso coffee or strong filter coffee (see introduction above)

Line a deep (2–3cm) tray, baking dish or container, about 20cm square, with greaseproof paper.

If you don't have a sugar thermometer, have ready a bowl of iced water.

Put the chocolate into a heatproof bowl over a pan of simmering water (make sure the base doesn't touch the water) and let the chocolate melt.

Put the cream, sugar, syrup and butter into a pan and bring to the boil, stirring continuously until you reach the 'soft ball' stage. This will be at 115°C if you have a sugar thermometer. If not, to check it is ready, dribble a little of the mixture into the iced water, give it 10 seconds or so to cool, then, with your fingers, see if you can roll it into a small ball. If you can, it is ready.

As soon as it reaches this point, take the pan from the heat and carefully stir in the melted chocolate, along with the almonds and the coffee. Beat slightly with a spatula or wooden spoon until the mixture binds together.

Pour into the lined tray and allow to cool, then put into the fridge to harden for 24 hours before cutting into squares.

Variation: White Chocolate and Cranberry Fudge

This fudge was first created one Christmas when we decided to make a festive-looking fudge that people might like to give as a present, but it was so popular it has stayed constantly in the farmshop ever since.

It is made using the same method, but with different quantities.

To make 500g fudge, line a deep 20cm x 10cm tray, baking dish or container, as in the previous recipe, and melt 150g white chocolate in the same way. Follow the rest of the instructions, but use 200ml double cream, 200g caster sugar, 20g golden syrup and 30g unsalted butter.

As soon as the mixture reaches 'soft ball' stage, take the pan from the heat and carefully stir in 80g dried cranberries and the melted chocolate and continue with the recipe.

Double Chocolate Chip Cookies

These are at their best straight from the oven – and even after you have made them and they are cold, you can put them into the oven briefly just to warm them through. They stay quite rustic-looking and don't have any real crunch, but are soft and melting in the middle.

250g butter, softened	400g plain flour
175g caster sugar	2 teaspoons baking powder
160g light brown sugar	60g cocoa powder
1 large egg, beaten	250g white chocolate buttons

Preheat the oven to 180°C/gas 4 and line four baking trays with greaseproof paper (alternatively bake the cookies in batches).

Put the butter and sugars into a bowl and cream together until pale and fluffy, then beat in the egg.

Sift together the flour, baking powder and cocoa powder and fold into the butter mixture until you have a soft dough. Mix in the white chocolate buttons.

Take pieces of the mixture and roll into balls the size of a golf ball (you should have enough for 20). Place well apart on the prepared baking trays, as the mixture will spread a little during cooking, and squash and flatten them slightly.

Put into the oven and bake, in batches if necessary, for about 12 minutes, until firm on the outside but soft in the middle. Take out of the oven and leave to cool on the trays.

STAPLES

Notes on jars and bottles

This chapter includes a collection of fresh dressings, pestos, mayonnaise and sauces that can be kept in the fridge: bottled sauces for your storecupboard and chutneys, pickles and jams that make the most of seasonal fruit and vegetables when they are at their prime, capturing their flavours for another day.

To sterilise your preserving jars or bottles, place them, open, on a tray in the oven at 110°C/gas ¼ for 10 minutes. Fill them, while still hot, with hot chutney, jam or sauce, then close the jars or bottles.

If you are only making a small quantity to be eaten in a matter of weeks, then this is all you need do, but if you are making use of a glut of fruit or vegetables and making larger quantities to keep, it is best to have the jars or bottles good and airtight, to stop any air getting in and mould growing. If you happen to have a steam function on your oven, as we do in our kitchens, put the closed jars in for 20 minutes to create a vacuum and seal them. Take out using a tea towel, and retighten the lid on each jar. Otherwise, boil the jars on the hob. Line a large pan with a cloth or tea towel at the base, which will stop the glass of the jars cracking, then put in your jars so that they aren't touching one another. Pour in enough boiling water to cover the jars completely, then put a lid on the pan and simmer for 10 minutes. Leave them in the pan until cool enough to touch, or lift out using tongs, again retightening the lids (if using screw caps) as they come out, then dry the jars, label and store.

French Dressing

3 tablespoons white wine vinegar

juice of 1½ lemons

1 tablespoon honey

3 tablespoons Dijon mustard

200ml grapeseed or vegetable oil

200ml extra virgin olive oil

sea salt and freshly ground black pepper

In a bowl, whisk together the vinegar, lemon juice, honey and mustard until blended. Combine the oils in a jug and very slowly start adding to the bowl: just a few drops at first, then work up to a slow, steady trickle, whisking constantly until all the oil is incorporated and the dressing will coat the back of a spoon. Season with salt and lots of freshly ground black pepper. Put into the fridge and chill.

Mayonnaise

2 egg yolks

1½ tablespoons Dijon mustard

1 tablespoon white wine vinegar

juice of ¼ of a lemon

500ml grapeseed or vegetable oil

sea salt and freshly ground black pepper

In a bowl, whisk together the egg yolks, mustard, vinegar and lemon juice until blended. Very slowly start adding the oil: just a few drops at first, then work up to a slow, steady trickle, whisking constantly until all the oil is incorporated and you have a thick, creamy mayonnaise. If you add the oil too quickly the mayonnaise will split, but if this happens, you can rescue it by putting a tablespoon of water into a clean bowl and then whisking in the split mixture, drop by drop, until properly emulsified.

Season with salt and lots of freshly ground black pepper. Chill until ready to use. It will keep in the fridge for up to a week.

Salsa Verde Mayonnaise

MAKES 500ML

This is an amalgamation of two classics: mayonnaise and salsa verde, the famous Italian green sauce, full of herbs and made piquant with capers and anchovies. You could add the flavourings to a good bought mayonnaise or see page 335 if you want to make your own. Serve it as a dipping sauce with fresh vegetable crudités – when we do this we also put out a pot of hot Cheddar sauce (see page 341) for a real contrast of temperatures and flavours.

Once made, this will keep for up to a week in the fridge.

- 40g fresh parsley leaves, roughly chopped
- 20g fresh mint leaves, roughly chopped
- 30g rocket leaves, roughly chopped
- 1 clove of garlic, roughly chopped
- 3 tablespoons capers
- 4 heaped teaspoons Dijon mustard
- 160ml olive oil
- 4 anchovy fillets
- 160ml mayonnaise

Mix all the ingredients apart from the mayonnaise in a bowl and leave to stand for 10 minutes to allow the flavours to infuse into one another. Blend to a coarse paste (or do this with a pestle and mortar).

Put the mayonnaise into another bowl and slowly whisk in the blended herb mixture, a little at a time. Do this slowly, to avoid the mayonnaise separating. Chill in the fridge until needed.

MAKES 150ML / MAKES 300ML

Herb Mayonnaise

A simpler herb mayonnaise, which is good with steamed vegetables, chicken or fish.

- 1 heaped teaspoon chopped fresh parsley
- 1 heaped teaspoon chopped fresh tarragon
- 2 heaped teaspoons chopped fresh mint
- 1 heaped teaspoon chopped fresh chives
- 125ml mayonnaise (see page 335)
- sea salt and freshly ground black pepper

Mix the chopped herbs into the mayonnaise, taste and season as necessary, then chill in the fridge until needed.

Watercress or Wild Garlic and Pumpkin Seed Pesto

Great spooned into a bowl of soup or with grilled fish or meat.

- 130g watercress leaves or sliced wild garlic leaves
- 1 clove of garlic, thinly sliced
- 130ml olive oil
- 3 tablespoons pumpkin seeds, lightly toasted in a dry pan
- 25g Parmesan cheese, grated
- sea salt and freshly ground black pepper

In a bowl, mix half the watercress or wild garlic with the garlic clove and olive oil and leave to stand for 10 minutes, then transfer to a blender and whizz to a smooth paste. Add the remaining watercress or wild garlic, the pumpkin seeds and Parmesan and pulse for a few seconds until you have a coarse pesto. Taste and season as necessary. Put into the fridge to chill (for up to a week) if not using straightaway.

MAKES 600ML

Minted Aioli

The classic garlic mayonnaise is freshened up with loads of mint, so is great with roast lamb, and especially the pressed lamb on page 219.

It is at its best chilled in the fridge for 24 hours, so that the flavour of the mint can really infuse all the way through.

4 egg yolks

2 cloves of garlic, crushed

juice of 1 lemon, strained

250ml olive oil

250ml rapeseed oil

4 tablespoons chopped fresh mint leaves

sea salt and freshly ground black pepper

In a bowl, whisk together the egg yolks, garlic, lemon juice, sea salt and freshly ground black pepper.

Very slowly, start adding the oil: just a few drops at first, then, as the mixture starts to thicken, work up to a slow, steady trickle, whisking constantly until all the oil is incorporated and the aioli is thick and creamy. Add a little water (1–2 tablespoons) if necessary to get the right consistency, then gently stir in the chopped mint. Taste and adjust the seasoning as necessary. Spoon into a bowl and, ideally, leave in the fridge for 24 hours before serving.

Hot Cheddar Sauce

We put pots of this out, together with some salsa verde mayonnaise (see page 336), whenever we serve crudités from the garden – but it is also great with good bread for dipping into it, at the start of a meal. The sauce will keep for up to a week in the fridge.

2 egg yolks
300g crème fraîche
60g Cheddar cheese, finely grated
40g Parmesan cheese, finely grated
sea salt and freshly ground black pepper

Put the egg yolks and crème fraîche into a heatproof bowl over a pan of simmering water (making sure the base of the bowl doesn't touch the water) and whisk constantly until the mixture becomes hot and slightly thick. Take off the heat, add the grated cheeses, gently mix until all the cheese has melted (this will thicken the sauce fully), then check for seasoning.

If not using immediately, pour into a small container and leave to cool, then put into the fridge.

When you want to use it, reheat it very gently over a low heat, stirring constantly until just hot. Take care not to let it boil or the sauce will curdle and have the consistency of scrambled egg.

MAKES 2 X 1 LITRE JARS

Pickled Vegetables

When we have a glut of different vegetables, we often pickle them together, just poaching them in the pickling liquid, so they retain their essential flavour, look very pretty in the jar and are lovely just run through a salad or with cold meats. You can vary the vegetables, but choose mainly crunchy ones, with a few slightly softer ones, so that you have an interesting mix of textures. You can add small onions if you like – put them in at the same time as the fennel.

Stored in a cool, dry place, these will keep for up to 6 months.

Selection of vegetables, for example:

1 fennel, cut into chunks

2 carrots, sliced at an angle

1 small cauliflower, cut into florets

2 celery stalks, sliced at an angle

1 red pepper, deseeded and sliced

1 yellow pepper, deseeded and sliced

2 courgettes, sliced

For the pickling liquid:

500ml white wine

250ml white wine vinegar

100ml olive oil

115g caster sugar, preferably unrefined

2 tablespoons salt

3 cloves of garlic, halved

5cm piece of fresh root ginger, chopped

2 sprigs of fresh rosemary

2 sprigs of fresh thyme

1 bay leaf

1 teaspoon black peppercorns

½ teaspoon pink peppercorns

4 whole cloves

1 teaspoon fennel seeds

1 teaspoon coriander seeds

½ teaspoon juniper berries

1 star anise

First make the pickling liquid. Put all the ingredients into a large pan with 1.5 litres of water, bring to the boil, then turn down to a simmer for 5 minutes. Take off the heat and leave to infuse for 3 hours, then remove all the halved garlic cloves and discard.

Put the pan back on the heat and bring back to the boil, then add the vegetables, a variety at a time, at 1-minute intervals, starting with the crunchiest ones (in this case the fennel) and finishing with the softest (the courgette), so that each one retains its crunch or bite.

Have ready your hot, sterilised jars (see page 334). Lift out the hot vegetables with a slotted spoon and divide between the jars, then top up with the hot pickling liquid (keep in the herbs and spices). Close the jars, then seal in a pan of boiling water (see page 334).

Variation: Piccalilli

We also pickle vegetables in a classic piccalilli. We make ours with chopped shallots and onions, cauliflower and cucumber, but you can add broccoli, green beans, peppers or whatever you like, so long as it has a good crunch.

For about 2kg, cut 1 large cauliflower into small florets, chop 2 medium white onions small (about 1cm), and cut 350g small shallots into quarters. The key is to draw as much moisture out of the vegetables as possible, so that the piccalilli doesn't become too liquid, so put the cauliflower, onions and shallots into a bowl with 3 teaspoons of sea salt and leave in the fridge for 24 hours, then drain in a colander under the cold tap, to wash off the salt, and leave to 'drip dry'. Deseed and chop ½ cucumber (again about 1cm), repeat the salting process with another 3 teaspoons, but for just 20 minutes, drain, rinse and leave to dry.

To make the sauce, put 350ml white wine vinegar and ½ a chopped red chilli (with seeds) into a medium, heavy-based pan and bring to the boil, then turn off the heat, leave to infuse for 20 minutes, and pass through a fine sieve to remove the chilli (keep the pan as you will need it again shortly). Next, put 250g caster sugar into a bowl with 25g English mustard powder, 1 tablespoon ground turmeric and 1½ tablespoons cornflour and whisk in half the infused vinegar, smoothing out any lumps.

Pour the remaining infused vinegar into the reserved pan and bring to the boil, then add the vinegar and flour mixture, bring back to the boil and whisk until the mixture starts to thicken, taking care not to let it get lumpy. Take it off the heat and, while still hot, pour over the vegetables.

Have ready your hot, sterilised jars (see page 334), fill them with the hot piccalilli, close the jars, then seal in a pan of boiling water (see page 334).

MAKES ABOUT 2.5KG

Chutneys

For me a good chutney is about the balance of sweet and sour flavours: you don't want it to be overpoweringly vinegary or sugary because you should be able to taste all the key ingredients.

Use a stainless steel pan, which won't react with the acidity in the vinegar (and could give a metallic taste to the chutney), and stir regularly during the cooking, so that the chutney doesn't catch and burn on the bottom of the pan.

Once potted in sterilised and sealed jars, all these chutneys will keep for a couple of years and become richer and more mellow as they mature. As with any preserves, though, use your eyes and nose, and if you think anything is wrong when you open the jar, discard the contents; however, it is very rare to have a problem with a chutney, as the very nature of it is that it is high in acidity (from the vinegar) and sugar – two things that bacteria don't like.

Apple and Chilli Chutney

This recipe dates back to the opening of the farmshop back in 2002, when John Hardwick and Tom Aikens between them made 22,000 jars of jams, pickles and chutneys from a standing start, in small batches with just a single pan on the stove. This one works beautifully in a ploughman's-style sandwich, made with good bread, Cheddar and ham, and is equally good with pork and terrines.

Over the years it has been made with many different varieties of apple from the orchards – some cookers, some eating apples, some heritage varieties – whatever has tasted right at the time of making. Cookers are easiest to get hold of at home, but if you want to experiment with varieties other than Bramleys, essentially you want an apple that will break down a little, not stay intact, but will give some good acidity, as a cooking apple does. Jez suggests Newton Wonder, Howgate Wonder or Blenheim Orange, the classic dual-purpose apple which is quite tart around the end of September to October but sweetens later. These won't soften quite as much as Bramleys, which are much bigger. The rule of thumb with apples is that the bigger the fruit and the larger its cells, the more it will disintegrate.

The chilli just gives a background, fragrant spiciness, rather than a kick of heat, as you don't want it to dominate the other ingredients. Generally the smaller the chillies, the hotter they are, so chose a medium-sized one. That said, all chillies vary in their heat levels, so taste a little raw before you start making the chutney, and then you can use more or less, accordingly.

2kg cooking apples, peeled, cored and chopped

350g sultanas

750g demerara sugar

50g yellow mustard seeds

2 red chillies, deseeded and chopped

2 tablespoons ground ginger

3 medium red onions, thinly sliced

6 cloves of garlic, finely chopped

500ml white wine vinegar

Put all the ingredients into a heavy-based pan with 250ml water and bring to the boil, stirring frequently to prevent sticking on the bottom of the pan, then turn down the heat and simmer for 1–1½ hours – again stirring regularly – until the mixture is soft and most of the liquid has evaporated. Take off the heat.

Have ready your hot, sterilised jars (see page 334), fill them with the hot chutney, close the jars, then seal in a pan of boiling water (see page 334).

Butternut Squash Chutney

Choose a butternut squash with a really bright orange flesh, rather than a pale one, if possible, as otherwise the tomatoes, brown sugar and raisins will dull down the colour. Also look for a tight flesh, not loose and broken, as this way you will get a bit more texture to your chutney.

2 large butternut squash (about 1.5 kg), peeled, seeds removed, flesh finely chopped

500g tomatoes, roughly chopped

500g onions, finely chopped

500g cooking apples, finely chopped

500ml cider vinegar

500g soft brown sugar

250g raisins

1 teaspoon mixed spice

2 teaspoons sea salt

½ teaspoon black pepper

Put the squash, tomatoes, onions, apples, vinegar and all the rest of the ingredients into a heavy-based pan with 250ml of water. Bring to the boil, stirring frequently to prevent sticking on the bottom of the pan, then turn down the heat and simmer for 1½ hours – again stirring regularly – until the mixture has thickened and there is only a little liquid left in the pan. Take off the heat.

Have ready your hot, sterilised jars (see page 334), fill them with the hot chutney, close the jars, then seal in a pan of boiling water (see page 334).

MAKES 1.5KG

Red Tomato Chilli Jam

This is a good way of using up ripe tomatoes that are becoming too soft to use for salads. You cook the jam quite quickly so it retains a beautiful colour. The chilli flavour works best with something pungent like a mature Cheddar or a goat's cheese.

1kg ripe red tomatoes, roughly chopped

2 red chillies, finely chopped (keep in the seeds)

1kg caster sugar

1 bay leaf

1 teaspoon cumin seeds

1 teaspoon fennel seeds

1 teaspoon black peppercorns

½ teaspoon cayenne pepper

2.5cm piece of fresh root ginger, very finely chopped

3 cloves of garlic, finely chopped

2 teaspoons salt

juice of 1½ lemons

650ml red wine vinegar

Put the tomatoes and chillies into a bowl, cover with the sugar, stir and leave in the fridge overnight.

When ready to make the jam, put a saucer into the fridge and get it really cold, ready to test the setting point. Make a little spice bag by laying the bay leaf in the centre of a small square of muslin, add the cumin and fennel seeds, peppercorns and cayenne pepper, then gather up and tie securely.

Transfer the tomato and chilli mixture to a heavy-based pan, with the ginger and garlic, salt, lemon juice and vinegar, and add the spice bag.

Bring to the boil, then turn down the heat to a simmer, stirring regularly to ensure the mixture doesn't burn on the bottom of the pan. After about 30 minutes, test to see if the jam has reached setting point. Take the pan off the heat and the cold saucer from the fridge. Spoon out a tablespoon of jam on to it, wait a minute, then push your finger through the middle of it. If it leaves a line that stays clean (i.e. the jam doesn't run back into it) and the jam wrinkles slightly, it has reached setting point. If not, continue to boil, and keep testing. Have ready your hot, sterilised jars (see page 334), fill them with the hot jam, close the jars, then seal in a pan of boiling water (see page 334).

Notes on cooking at Christmas

As countless chefs and home cooks will tell you, cooking around the festive season is made a lot less stressful with preparation. Lists and schedules can really help. At home, we start our preparations on stir-up Sunday, when we'll make our Christmas pudding. Then the cake will follow. Every household is different and what's traditional in your house could be very different from mine, so on the following pages our chefs have included some recipes and advice for the Christmas classics.

Christmas Turkey: Richard Smith reckons he gets more favourable feedback on our turkeys than practically anything else on the farm. The secret is that they are finished on oats, which are an oil-based cereal, and so not only do the turkeys go mad for them, but they give the finished meat a superb quality.

With turkeys, because they are so big, John Hardwick breaks with his advice about not being too guided by charts. It is important to weigh the bird, as you don't want to have everyone sitting around the table and the rest of the meal ruining because you have seriously miscalculated the cooking time.

Preheat the oven to 200°C/Gas 6. John cooks any stuffing separately, as he believes that stuffing the cavity benefits the turkey more than the stuffing. Smear the turkey with olive oil and butter, and also grease the roasting tin with a little oil, too. Season the skin with sea salt and pepper and rub it in well. Put the turkey in the oven and cook for 40 minutes (skin side down), then turn down the heat to 180°C/gas 4 and cook for 35 minutes per kg. Keep basting as the turkey cooks, and as the juices start to form in the roasting pan, add about 300ml of water to them, and top this up every so often, to create moisture and also good juices for your gravy.

After about 2 hours, carefully turn the bird back over, so the breasts are facing upwards, and finish cooking. If the breasts start to colour too much, cover in foil. To check that the bird is ready, insert a skewer into the thickest part of the thigh – if the juices run clear, it is cooked. Take out of the oven, cover with foil and allow to rest for 30 minutes before serving.

Aunt Mimi's Mince Pies

MAKES 24

Aunt Mimi is my mother-in-law's sister and she is a wonderful baker. Whenever I made mince pies at home, Anthony, my husband, would always say: 'Not quite as good as Auntie Mimi's'.

Of course you can buy good mincemeat – but homemade is what makes the pies special. You will need to leave it to marinate in an airtight container for at least a week, but it can keep for up to 1 month. Using lard in the pastry gives it a crisper finish but you can substitute it for butter, increasing the total butter quantity to 100g, if you prefer.

If you can find individual candied orange and lemon peel, then use a ratio of two-thirds orange to one-third lemon, as this makes the mincemeat more floral than sharp. Any leftover pastry can be used to make a couple more pies or wrap, label or freeze it for future use.

For the mincemeat:

115g sultanas

115g raisins

70g currants

2 tablespoons flaked almonds

2 tablespoons dried cherries

115g crystallised ginger

35g mixed candied peel (see above)

500g grated Bramley apples (about 3)

85g soft light brown sugar

4 tablespoons Cognac

zest of 3 lemons

zest of 3 oranges

a pinch of mixed spice

a pinch of ground nutmeg

70g lard, grated, or suet or butter

a pinch of sea salt

For the pastry:

200g plain flour, plus extra for rolling out

70g cold butter, grated, plus a little extra for greasing

30g cold lard, grated, or butter

50g caster sugar

a pinch of salt

1 medium egg, beaten, plus 1 egg, beaten, for glazing (optional)

To make the mincemeat, mix together all the ingredients and allow to marinate in a lidded, airtight container in the fridge for 1 week.

Lightly butter a 12-hole tart tray.

Put the flour, butter and lard into a bowl and rub with your fingertips until the mixture resembles breadcrumbs. Stir in the sugar and salt, then gradually add the beaten egg until combined and the mixture comes together in a dough. If the mixture is dry, stir in 1–2 tablespoons of iced water and continue to gently knead it together. Form into a ball, flatten into a disc (it makes it easier to roll out), wrap in clingfilm, and chill in the fridge for 1 hour before using. This is very a very soft pastry, so it needs this quite long chilling time to firm up and make it easier to roll.

Lightly flour your work surface. Cut the ball of dough in half and keep the second ball in the fridge while you roll out the first one to about 4mm thick. Using a plain round cutter, 7cm in diameter, cut out 12 circles and press into the tart tray – take the time to do this gently and carefully, so that the pastry doesn't crack. Spoon 1 full tablespoon of mincemeat into each tart, making sure it domes slightly in the centre.

Roll out the remaining pastry and cut out as above. Gently lay a circle of pastry over the top of each mincemeat-topped base and gently crimp the edges together with the end of a small round table knife all the way round to seal. Brush with a little beaten egg, if you like, then insert the tip of a knife into the middle of each pastry top to make a tiny slit for steam to escape. Put the tin into the fridge to rest again for 1 hour.

When ready to bake, preheat the oven to 190°C/gas 5.

Remove the tray from the fridge and bake in the oven for 25–30 minutes, until the pastry is golden brown. Take out of the oven and leave to cool in the tray for 5 minutes before removing and cooling on a wire rack.

MAKES 1 X 18CM ROUND CAKE

Christmas Cake

The key to keeping this cake moist is to make it a few weeks or even months in advance and 'feed' it regularly by making holes in it with a skewer, spooning over a little brandy and allowing the alcohol to seep through the cake. However, there is no need to panic if you leave it to the last minute: just make sure that the fruit is soaked well in brandy, and properly plumped up before you start to make your cake. You can leave it to soak for up to 3 days if you like: the longer the better.

If you are icing the cake, you can do this a couple of weeks before Christmas.

225g raisins	170g soft light brown sugar
225g currants	1 tablespoon treacle or molasses
225g sultanas	1 teaspoon vanilla extract
60g dried cherries	3 medium eggs
35g toasted flaked almonds	170g plain flour
4 tablespoons mixed candied peel	½ teaspoon mixed spice
zest and juice of 1 lemon	½ teaspoon ground ginger
zest and juice of 1 orange	½ teaspoon ground nutmeg
100ml brandy, plus extra for feeding	½ teaspoon ground cinnamon
170g butter, softened, plus a little for greasing the tin	½ teaspoon ground cardamom

Put all the dried fruit, nuts, peel, citrus zest and juice into a bowl with the brandy, mix well, then cover (with muslin, ideally) and leave to soak overnight.

The next day, preheat the oven to 160°C/gas 3. Grease an 18cm round tin (with a removable base) with butter and line it with a double layer of greaseproof paper.

Cream together the softened butter and sugar in a mixing bowl until pale and fluffy, then gradually beat in the treacle or molasses, vanilla extract and eggs. When all is incorporated, add the soaked fruit, flour and spices and gently fold in. Spoon into the prepared cake tin and smooth the top. Put into the oven and bake for about 2½–3 hours, or until a skewer inserted into the centre comes out clean.

While the cake is still warm, pierce with the skewer and 'feed' with brandy, two or three times (more if you like) as the cake matures.

Christmas Pudding

MAKES 1 X 1.1 LITRE PUDDING

This is my mother's recipe and is the one we make every year at home.

120g suet
60g self-raising flour, sifted
120g white breadcrumbs
240g demerara sugar
120g sultanas
120g raisins
120g currants
1 apple, grated
grated zest of 1 orange
grated zest of 1 lemon
1 tablespoon candied orange peel
1 tablespoon candied lemon peel
1 teaspoon mixed spice
1 teaspoon ground nutmeg
1 teaspoon ground cinnamon
30g chopped almonds
2 eggs
50ml rum
70ml white wine
70ml stout

In a large bowl, mix together the suet, flour, breadcrumbs and sugar, then add the dried fruits, apple, zests, peel, spices and almonds.

Beat the eggs with the rum, wine and stout in a separate bowl, then pour in and mix thoroughly to give a thick batter. Cover the bowl with clingfilm and leave overnight in the fridge. The next day, take it out and stir.

Grease a 1.1 litre ovenproof pudding basin with butter, fill the basin with the pudding mixture and smooth the top. Cover the basin with a double layer of greaseproof paper, make a pleat in the centre and then follow with a sheet of foil. Tie a piece of string around the basin to secure the foil and paper in place. Place on a trivet inside a large pan. Pour in enough boiling water to come halfway up the outside of the bowl, then cover the pan and keep the water simmering for 8 hours, topping up with boiling water from a kettle from time to time.

The pudding can either be eaten straightaway, or cooled, rewrapped and stored for up to a year. Re-steam in the same way for 2½ hours before serving.

Notes on tipples

Surrounded as we are at Daylesford by fruit from hedgerow to orchard, we can't resist using it in drinks as well as in our cooking, baking, bread-making and dairy produce. There is nothing better on a hot day than a glass of fresh lemonade or a fruit and wine cup, served from a big glass jug or bowl when friends come around, and we like to keep up the old English tradition of seasonal fruit liqueurs. Even when we put water carafes on the table, we like to fill them with long strips of cucumber or a mix of cucumber, lemon slices and mint.

Lemon Refresher

This is a proper old-fashioned recipe. The Epsom salts and acids can be found in any chemist, and they are there to make a few bubbles and zing, in contrast to the more usual flat, homemade lemonade.

Peel the rind from 5 large unwaxed lemons with a vegetable peeler and put into a large earthenware bowl. Cut the lemons in half and squeeze the juice into the bowl, too. Add 800g caster sugar, 25g Epsom salts, 12g citric acid and 8g tartaric acid. Measure 1 litre of boiling water and pour over the top, then stir with a wooden spoon until all the sugar has dissolved. Cover the bowl with clingfilm and leave overnight in the fridge.

The next day, have ready two sterilised litre bottles (see page 334). Strain the liquid through a conical sieve into the bottles, seal and keep in the fridge (for up to a month). Serve over ice in tall glasses garnished with sprigs of fresh thyme.

Daylesford Summer Wine Cup

For a jugful big enough for about 12 glasses, you need 1 punnet of strawberries, stalks removed and halved, plus ½ a pineapple, skin cut off, 2 ripe peaches, stoned, 2 apples, cored and 1 mango, peeled – all of these need to be chopped quite small (about 1cm). You also need a few extra halved strawberries and some more neatly cut, with some small pieces of the other fruit to garnish, plus about 12 sprigs of mint.

The Summer Wine Cup needs a little sugar syrup to sweeten it, but how much you put in is really up to you. The best thing to do is make up more than you need and then add a little at a time, until you are happy with the depth of sweetness. We make our syrup in the ratio of one part sugar to two parts water.

Make the sugar syrup first by putting 12 tablespoons sugar into a pan with 360ml of water and heating, stirring, until the sugar has dissolved and you have a clear syrup. Take off the heat and leave to cool.

Put all the fruit into a large bowl, then pour in 2 bottles of dry white wine, 2 tablespoons each of Maraschino (cherry) liqueur and Grand Marnier, and add your sugar syrup to taste. Put into the fridge for 48 hours, to marinate, then strain through a fine sieve into a large glass jug containing some ice. Add the fruit garnish and mint sprigs.

Seasonal Liqueurs

We make Blackcurrant Vodka, Plum Brandy and Sloe Gin as the fruit becomes available. For the sloe gin, the best time to pick the fruit is after the first frost, as the sloes will be a lot sweeter and will have more flavour.

All the liqueurs are made using the same method. We divide a litre of the appropriate alcohol between two large (2 litre) sterilised kilner jars (see page 334), together with the fruit (350g blackcurrants, washed and stalks removed, for the vodka; 660g plums or damsons for the brandy and 450g sloes, washed and pricked, for the gin) and granulated sugar (200g for the vodka, 330g for the brandy and 210g for the gin).

Then we seal the jars and leave them to stand in a cool, dark place (for example a garage) for 10 months, shaking the jars every month for the first 4 months to fully dissolve the sugar. After 10 months, we strain the liqueur through three layers of muslin cloth into a jug, then pour through a funnel into a sterilised bottle or bottles and seal, making sure this is airtight. The liqueur is now ready for drinking.

MAKES ABOUT 12 SMALL BOTTLES

Two Bottled Sauces

The nation's favourites – even better if you make them yourself.

Tomato Ketchup

1 teaspoon black peppercorns
1 teaspoon whole allspice berries
½ teaspoon cloves
½ teaspoon cayenne pepper
3.5kg tomatoes, roughly chopped
900g onions, peeled and finely chopped
1kg cooking apples, peeled, cored and chopped
800g caster sugar, unrefined
1 litre malt vinegar
1 tablespoon sea salt

Make a little spice bag by heaping the peppercorns, allspice, cloves and cayenne pepper in the centre of a small square of muslin, then gather up and tie securely.

Put the tomatoes, onions, apples and sugar into a large heavy-based pan with the muslin spice bag. Add the vinegar and the salt. Bring to the boil, then turn down to a simmer for about 2 hours, stirring from time to time to make sure the mixture doesn't stick, until you have a pulp.

Remove from the heat and leave to cool a little. Purée in a blender, a little at a time, until smooth. Strain through a medium sieve into a clean pan and bring back to the boil for 2–3 minutes, stirring constantly to prevent catching and burning.

Have ready your hot, sterilised bottles (see method for jars on page 334). Pour the ketchup, through a funnel, into the bottles and close up, then seal in a pan of boiling water (see page 334).

Brown Sauce

MAKES ABOUT 12 SMALL BOTTLES

'This is a great sauce,' says John Hardwick. 'The recipe comes from Ivan, our sous chef in the farmshop, and we had to prise it from him. It gives HP a run for its money and is great with corned beef hash (see page 41).'

Jez suggests Blenheim Orange, Bountiful and the English Golden Delicious apples. 'People have a blinkered view of this apple because of the flabby French variety of the seventies and eighties,' he says. 'Whereas the English variety is a marvellous apple – hard, dense and sweet.'

- 250g pitted prunes
- 850ml malt vinegar
- 50g sea salt
- 1 teaspoon cayenne pepper
- 1 tablespoon ground allspice
- 2 teaspoons ground ginger
- 2 teaspoons grated nutmeg
- 750g (about 4 medium) onions, chopped
- 1kg apples (see above), peeled and chopped
- 500g caster sugar, preferably unrefined

Put all the ingredients into a large, heavy-based pan and bring to the boil, then turn down to a simmer for about 2 hours, stirring from time to time to make sure the mixture doesn't stick, until you have a pulp.

Remove from the heat and leave to cool a little. Purée in a blender, a little at a time, until smooth. Strain through a medium sieve into a clean pan and bring back to the boil for 2–3 minutes, stirring constantly to prevent catching and burning.

Have ready your hot, sterilised bottles (see method for jars on page 334). Pour the sauce through a funnel into the bottles and close up, then seal in a pan of boiling water (see page 334).

Two jams, a conserve and a marmalade

All our jams are set naturally without any added pectin, so we rely on what is naturally in the fruit and on the method of reducing the mixture of fruit, sugar and water right down until there is no liquid left, by which time the jam should have reached setting point. The downside is that there is a small window between the jam retaining its bright colour and looking darker. You also need to watch it all the time to make sure it doesn't catch and burn. We have tried all kinds of traditional methods to add pectin, including blitzing together lemon pith and flesh and adding it to the jam pan in a muslin bag or asking the bakery to keep back all the apple skins after they have made spiced apple cakes and adding these too, in muslin bags, but the reality is, you need masses of all of these trimmings to make a real difference.

As a rule, the more tart the fruit, the more pectin it contains, which is why we have chosen most of the fruits in the recipes that follow. And because they are quite tart to begin with they can cope with the concentrated sweetness that comes from reducing the jam right down. The exception is strawberry, which we make as a conserve, rather than a jam – as it demands less sugar, and has a slightly softer set.

Rhubarb and Ginger Jam

The pairing of warming, slightly fiery ginger with tart rhubarb is one of my favourites and works well if you like a jam with a little bit of a kick. Rhubarb can be quite low in pectin and tends to break down into a purée as it cooks, so the texture is a little different to other jams, but the flavour is wonderful.

1.6kg rhubarb, washed and cut into 2.5cm pieces

1.3kg granulated sugar

juice of 2 lemons

25g fresh root ginger, very finely chopped

The day before you want to make the jam, put the rhubarb, sugar and lemon juice into a bowl and leave to macerate in the fridge overnight. Just before cooking, stir in the chopped ginger.

When ready to make the jam, put a saucer into the fridge and get it really cold, ready to test the setting point.

Transfer the macerated rhubarb into a large, heavy-based pan with 300ml of water and bring to the boil, skimming off any scum with a slotted spoon. Turn down to a simmer, stirring regularly to avoid catching and burning. Remove the scum as it appears.

After about 30 minutes, test to see if the jam has reached setting point. Take the pan off the heat, and the cold saucer from the fridge. Spoon out a tablespoon of jam on to it, wait a minute, then push your finger through the middle of it. If it leaves a line that stays clean (i.e. the jam doesn't run back into it) and the jam wrinkles slightly, it has reached setting point. If not, continue to boil, and keep testing.

Have ready your hot, sterilised jars (see page 334). Take the pan from the heat and leave the jam to stand for 5 minutes before filling and closing the jars, then seal in a pan of boiling water (see page 334).

Gooseberry and Elderflower Jam

MAKES ABOUT 2.5KG

The combination of gooseberries and elderflowers is an English classic. With gooseberries, the best are always mid-season: not too hard and sour, as the earliest can be, and not too soft and verging on the sweet, as in the late ones. Gooseberries are high in pectin, so it is easy to get this jam to set.

1.4kg gooseberries, topped and tailed	1.4kg granulated sugar
juice of 1 lemon	200ml elderflower cordial

Put a saucer into the fridge and get it really cold, ready to test the setting point of the jam.

Wash the gooseberries really well and put them into a large heavy-based pan over a medium heat with the lemon juice and 250ml of water. Cook gently, ensuring the sugar doesn't burn before the fruit has turned to pulp, stirring regularly to avoid sticking.

Add the sugar and cordial and bring to the boil, skimming off any scum with a slotted spoon, then turn down to a simmer and stir regularly to avoid catching and burning. After about 30 minutes test to see if it has reached setting point. Take the pan off the heat, and the cold saucer from the fridge. Spoon out a tablespoon of jam on to it, wait a minute, then push your finger through the middle of it. If it leaves a line that stays clean (i.e. the jam doesn't run back into it) and the jam wrinkles slightly, it has reached setting point. If not, continue to boil, and keep testing.

Have ready your hot, sterilised jars (see page 334). Take the pan from the heat and leave the jam to stand for 5 minutes before filling and closing the jars, then seal in a pan of boiling water (see page 334).

MAKES ABOUT 2.5KG

Strawberry and Vanilla Conserve

The difference between a jam and a conserve is the ratio of sugar to fruit. A jam must have a minimum of 65 per cent sugar, – fine for quite acidic fruit, like gooseberries, blackcurrants, rhubarb and even raspberries, but can be too sweet for delicate fruits like strawberries. So we have lowered the sugar slightly here to let the delicate flavours of the vanilla and strawberries come through and, because of this, we have to label it a conserve. The butter helps to stop the quite loose and syrupy watery mix from boiling over at the beginning of cooking.

1.8kg strawberries, washed well, hulled and stalks removed

1.4kg granulated sugar

juice of ½ lemon

1 small knob of butter

2 vanilla pods, split, each pod cut so that you have 1 piece per jar

Forty-eight hours before you want to make the conserve, put 500g of the strawberries into a bowl and mix with the sugar. Leave to macerate in the fridge.

When ready to make the conserve, put a saucer into the fridge and get it really cold, ready to test the setting point.

Put the rest of the strawberries into a large heavy-based pan with the lemon juice and 300ml water and cook gently for a couple of minutes.

Add the sugared berries and the butter, and bring to a rapid boil, skimming off any scum. Continue to boil (it is important to cook quickly to maintain the colour and stop the berries from breaking up), stirring regularly to avoid catching and burning. After about 5 minutes, put in the vanilla pods, scraping in the seeds. Then, after another 5–6 minutes, test to see if the conserve has reached setting point. Take the pan off the heat, and the saucer from the fridge. Spoon out a tablespoon of conserve on to it, wait a minute, then push your finger through the middle of it. If it leaves a line that stays clean (i.e. the conserve doesn't run back into it) and the conserve wrinkles slightly, it has reached setting point. If not, continue to boil, and keep testing.

Have ready your hot, sterilised jars (see page 334). Take the pan from the heat and leave the conserve to stand for 5 minutes before filling, making sure a piece of vanilla pod goes into each jar, and closing the jars, then seal in a pan of boiling water (see page 334).

Seville Orange Marmalade

MAKES ABOUT 1.5KG

This recipe works equally well with grapefruit. At Christmas we make a festive version with a glug of brandy or whisky added just before we take the marmalade off the heat. You can also add some mulled wine spices in a muslin bag during the cooking.

1.3kg Seville oranges, washed
1.3kg granulated sugar
juice of 1½ lemons

Put a saucer into the fridge and get it really cold, ready to test the setting point of the marmalade.

Put the whole washed oranges into a large heavy-based pan with 2 litres of water, bring to a simmer and cook gently until they are soft. Lift out the oranges from the pan with a spoon and cool slightly. Let the liquid in the pan bubble up, reduce by a third, then take off the heat, strain through a sieve into a bowl and pour back into the pan.

Top, tail and halve the cooled oranges and scoop out the flesh, keeping the peel and pith intact. Trim off all but a thin lining of pith, weigh the peel so that you have 150g, then shred into strips about 3mm wide.

Put the orange flesh, complete with pips, into a blender and blend to a purée. Strain through a fine sieve into the reduced liquid from cooking the oranges. Add the strips of peel, the sugar and the lemon juice and bring to the boil. Boil rapidly, stirring occasionally to avoid catching and burning on the bottom of the pan.

After about 30 minutes, test to see if the marmalade has reached setting point. Take the pan off the heat, and the cold saucer from the fridge. Spoon out a tablespoon of marmalade on to it, wait a minute, then push your finger through the middle of it. If it leaves a line that stays clean (i.e. the marmalade doesn't run back into it) and the marmalade wrinkles slightly, it has reached setting point. If not, continue to boil, and keep testing.

Have ready your hot, sterilised jars (see page 334). Take the pan from the heat and leave the marmalade to stand for 5 minutes before filling and closing the jars, then seal in a pan of boiling water (see page 334).

Notes on stocks

Never waste a vegetable or a bone: make stock! Making stock is so easy and so satisfying. It is really just about getting into the habit of automatically using up whatever vegetables you have left over in the fridge or saving the chicken or beef carcass after a roast, ready to boil up, rather than wasting it.

The key with meat stocks is to bring them to the boil, then turn down to a simmer and spend a good 10 minutes skimming off the 'scum', i.e. all the impurities and grease that rise to the surface, in what I call 'the first throw'. If you don't turn down the heat and leave the stock boiling at this point, the fats will emulsify into the stock. There is no real flavour in the stock yet, so you are taking nothing away in terms of taste, whereas if you leave it until later, you will be losing flavour. As the stock continues to simmer over a few hours, a tiny bit more scum will probably kick up to the surface, now and then, but you can skim that off easily without harming the flavour.

Once you have made your stock, you can keep it in the freezer, in bags or ice-cube trays, ready to pull out at any time for a soup or casserole, and it will make all the difference to the flavour. Above all, you will know exactly what is in your stock: nothing but pure goodness.

Vegetable Stock

You can vary the vegetables as much as you like, but avoid green vegetables, apart from leeks, as they tend to colour the water, which will eventually turn brownish in the way that it always will if you overcook greens and the stock will lose some of its fresh flavour.

To make about 3 litres, chop ¼ of a celeriac, 2 large carrots, 1 leek, 1 large onion, 3 sticks of celery, ½ a head of fennel, 1 sweet potato and ¼ of a butternut squash and put into a large pan. Cover with 5 litres of water and add 2 sprigs of fresh thyme, ½ teaspoon of sea salt and 1 teaspoon of black peppercorns. Bring to the boil, then turn down the heat and simmer for 1 hour. Take off the heat and leave to stand for 2 hours, to improve the flavour. Remove the vegetables from the pan with a slotted spoon and throw them away. Pass the liquid through a fine sieve into a clean container, using a ladle. Either keep in the fridge (for up to a week) or freeze.

Chicken Stock

In our kitchens, we tend to use chicken stock for most of our cooking, even if we are making a casserole of lamb or beef, because it is lighter and allows other flavours in a dish to come through more than, say, a beef stock. We make a white (light) chicken stock, which is good for soups and chicken casseroles, and a roast or 'brown' one, which we use in dark meat casseroles and for gravies and sauces – however, you can vary the latter recipe to use any meat bones, such as beef, lamb, duck or turkey.

White (light) chicken stock
To make 3.5 litres, put a 1.2kg chicken carcass into a large pan, cover with 4.5 litres of water and add 2 teaspoons sea salt. Bring slowly to the boil, then reduce the heat to a simmer and skim off the scum and grease from the surface with a slotted spoon. Add 2 peeled and quartered onions, 1 leek, washed well, split and halved, 3 halved sticks of celery, 6 cloves of garlic, 2 tablespoons chopped fresh thyme leaves and 1 tablespoon black peppercorns. Bring back to the boil, then turn down the heat and simmer for about 1 hour, carefully skimming off any more scum that rises to the surface from time to time.

Turn the heat off and leave to cool for 30 minutes. Remove the carcass and vegetables from the pot with a small sieve or slotted spoon and discard, then strain the liquid through a fine sieve into a clean container. Either put into the fridge until ready to use (it will keep for 2 days), or freeze.

Roast (brown) chicken stock
To make 3.5 litres, preheat the oven to 200°C/gas 6. Chop up a 1.2kg chicken carcass (or use the equivalent weight in chicken wings) and put into a roasting tin. Drizzle with 1 tablespoon of vegetable oil and put into the oven for about 30–40 minutes, or until golden brown.

Meanwhile, roughly chop 2 onions, 2 large carrots, 3 sticks of celery, and break a head of garlic into cloves. Heat 3 more tablespoons of vegetable oil in a large pan, then add the vegetables and cook over a medium heat, stirring occasionally, until they are golden brown, adding the garlic towards the end, so that it doesn't burn and become bitter. Add 2 tablespoons tomato purée and continue to cook for another 5 minutes, making sure to scrape the bottom of the pan or the tomato purée will burn.

Remove the roasting tin from the oven and, with tongs, add the roasted bones or wings to the pan and pour off any excess fat from the tin, keeping back the juices (you can keep the fat for the Sunday roast). Add a few tablespoons of water to the roasting tin and stir until all the roasting juices are released, then add these to the pan and cover with 4.5 litres of water. Bring to the boil, then reduce the heat to a simmer. Skim off the scum and grease from the surface with a slotted spoon, then add 4 sprigs of fresh thyme, a tablespoon of sea salt and a teaspoon of black peppercorns and continue to simmer for 2 hours, carefully skimming off any more scum that rises to the surface from time to time.

Take off the heat, allow to cool a little, then lift out the bones and vegetables with a small sieve or slotted spoon and discard. Strain the liquid through a fine sieve into a clean container. Either put into the fridge until ready to use (it will keep for 2 days) or freeze.

Weights and Measures

DRY MEASUREMENTS

Metric	US cups	Metric	Australian cups
150g	1 cup flour	125g	1 cup flour
225g	1 cup caster/granulated sugar	225g	1 cup caster/granulated sugar
175g	1 cup brown sugar	200g	1 cup brown sugar
225g	1 cup butter	250g	1 cup butter
200g	1 cup uncooked Arborio rice	220g	1 cup uncooked Arborio rice

WEIGHTS

Metric	Imperial
10g	½oz
25g	1oz
50g	2oz
75g	3oz
110g	4oz
225g	8oz
450g	1lb

LIQUID MEASUREMENTS

Metric	Imperial	US cups
30ml	1 fl oz	⅛ cup
60ml	2 fl oz	¼ cup
125ml	4 fl oz	½ cup
185ml	6 fl oz	¾ cup
250ml	8 fl oz	1 cup
500ml	16 fl oz	2 cups

Index

Adlestrop and kale tart **155**

Almonds
  apricot and almond tart **271**
  green beans with almonds, parsley and garlic butter **124**
  Jerusalem artichoke risotto with garlic and almond breadcrumbs **139–40**
  milk chocolate, almond and espresso fudge **329**
  Parmesan, chilli and Marcona almond biscuits **324**
  purple sprouting broccoli, spelt, crispy garlic and toasted almonds **105**

Apples
  apple and chilli chutney **346–7**
  blackberry and apple crumble tart **268**
  celeriac and apple soup **47**
  poached apple and pear jelly with crumble topping and prune cream **254–6**
  spiced apple cake with streusel topping **314**
  vanilla rice pudding with apple and blackberry compote **264**
  venison cottage pie with beetroot and apple salad **242–3**
  wild rice, red cabbage, apple and toasted cobnut salad **113**

Apricots
  apricot and almond tart **271**
  chicken and apricot curry **223**

Asparagus
  asparagus, spelt, peas and mint salad **76**
  garden vegetables with hot Cheddar sauce and salsa verde mayonnaise **15**
  goat's cheese and asparagus tart **156**
  pearl barley, asparagus and pea shoot risotto **145**

Aubergine, pomegranate, feta and pumpkin seed salad **94**

Bacon
  beetroot, bacon and crème fraîche soup **53**
  chicken, leek and bacon pie **172–3**
  Single Gloucester, spinach and smoked bacon tart **168**

Baywell cheese: red onion tarte tatins with Baywell cheese **160–61**

Beans
  broad bean, bulgar wheat and herb salad **73**
  broad bean, pea and watercress risotto **137**
  broad bean, pea, mint and feta toasts **16**
  green beans with almonds, parsley and garlic butter **124**
  lamb and butter bean casserole with tomatoes, caperberries and olives **213–14**
  raw beetroot, kidney beans and mustard leaf salad with horseradish dressing **106**
  slow-cooked lamb shoulder with white beans and salsa verde mayonnaise **215–16**
  smashed broad beans, peas and mint **121**
  smoky slow-cooked shin of beef chilli **232**
  spiced pumpkin, butter bean and spinach casserole **134**

Beef
  beef, ale and barley casserole, **238**
  braised brisket with lentils **234**
  corned beef **236–7**
  corned beef hash cakes **41**
  featherblade of beef with creamed wild mushrooms **230–31**
  roast rib of beef with Dijon mustard and balsamic sauce **227**
  smoky slow-cooked shin of beef chilli **232**

Beetroot
  beetroot, bacon and crème fraîche soup **53**
  beetroot, swede and potato bake **131**
  grilled mackerel with roasted beetroot and spiced lentils **185–6**
  raw beetroot, kidney beans and mustard leaf salad with horseradish dressing **106**
  raw slaw with chilli, soy and ginger dressing **85–6**
  venison cottage pie with beetroot and apple salad **242–3**

Biscuits
  blue cheese and walnut **323**
  Cheddar **322**
  double chocolate chip cookies **328**
  ginger **327**
  lemon shortbread **325**
  Parmesan, chilli and Marcona almond **324**

Blackberries
  blackberry and apple crumble tart **268**
  vanilla rice pudding with apple and blackberry compote **264**

Bledington Blue cheese and broccoli tart **159**

Blue cheese
  Bledington Blue cheese and broccoli tart **159**
  blue cheese and walnut biscuits **323**
  lentils, tomato, Daylesford Blue and red onion salad **99**
  pickled pear and hazelnuts with chickpeas, quinoa and Daylesford Blue **97–8**

Bread; notes on **282–3**
  baking with natural leaven **294–5**
  coffee jelly with brown bread ice cream **260–61**
  hot cross loaf or buns **289–91**
  nettle flowerpot bread **304–5**
  pumpernickel **286–8**
  red onion, Cheddar and chilli flowerpot bread **302–3**
  seven seeds sourdough, **299**
  sourdough **296–7**
  squash, honey and sage bread **284–5**
  tomato and sourdough salad with red pepper, onion and basil **82**
  see also toasts

Broccoli
  Bledington Blue cheese and broccoli tart **159**
  chestnut, quinoa, kale and broccoli salad **102–3**
  purple sprouting broccoli, spelt, crispy garlic and toasted almonds **105**

Brown sauce **361**

Brownies, dark and white chocolate **319**

Bubble and squeak with fried egg **35–7**

Bulgar wheat: broad bean, bulgar wheat and herb salad **73**

Buns, hot cross 289–91

Butternut squash
butternut squash and kale tart 163
butternut squash chutney 349
butternut squash, honey and sage soup 56
griddled butternut squash, goat's cheese and olive salad 93–4
squash, honey and sage bread 284–5

Cabbage
bubble and squeak with fried egg 35–7
crushed, buttered root vegetables and cabbage 127
raw slaw with chilli, soy and ginger dressing 85–6
wild rice, red cabbage, apple and toasted cobnut salad 113
see also cavolo nero; kale

Caesar salad, chicken 114–15

Cakes; notes on 308
blood orange and polenta with orange whipped cream 272–5
chocolate 309–11
Christmas 356
dark and white chocolate brownies 319
Earl Grey 315
lemon drizzle 317
Manuka honey 312
spiced apple with streusel topping 314

Carrots
crushed, buttered root vegetables and cabbage 127
raw slaw with chilli, soy and ginger dressing 85–6

Casseroles
beef, ale and barley 238
chicken with a splash of brandy 220–22
lamb and butter bean with tomatoes, caperberries and olives 213–14
spiced pumpkin, butter bean and spinach 134

Cauliflower: curried cauliflower, red pepper and nigella seeds 101

Cavolo nero
Jerusalem artichoke and cavolo nero tart 164
venison and cavolo nero lasagne 239

Celeriac
celeriac and apple soup 47
mushroom, celeriac, truffle honey and toasted pine nuts 110–11

Cheddar
Cheddar biscuits 322
Cheddar, potato and onion pie 169–71
garden vegetables with hot Cheddar sauce and salsa verde mayonnaise 15
hot Cheddar sauce 341
pan haggerty with mustard, egg and caper mayonnaise 39–40
red onion, Cheddar and chilli flowerpot bread 302–3
Welsh rarebit and chutney 18–19

Cheese, see under individual names

Chestnut, quinoa, kale and broccoli salad 102–3

Chicken
chicken and apricot curry 223
chicken Caesar 114–15
chicken casserole with a splash of brandy 220–22
chicken, ginger and vegetable broth 55
chicken, leek and bacon pie 172–3
chicken stock 373–4
ten vegetable and two grain minestrone 59

Chickpeas: pickled pear and hazelnuts with chickpeas, quinoa and Daylesford Blue 97–8

Chillies
apple and chilli chutney 346–7
garlic and chilli dressing 93
Parmesan, chilli and Marcona almond biscuits 324
raw slaw with chilli, soy and ginger dressing 85–6
red onion, Cheddar and chilli flowerpot bread 302–3
red tomato chilli jam 351
smoky slow-cooked shin of beef chilli 232
venison terrine with tomato and chilli jam 22–3

Chocolate
chocolate cake 309–11
dark and white chocolate brownies 319
double chocolate chip cookies 331
milk chocolate, almond and espresso fudge 329
white chocolate and cranberry fudge 330

Christmas; notes on cooking at 353
cake 356
pudding 357

Chutney 346
apple and chilli 346–7
butternut squash 349

Welsh rarebit and chutney 18–19

Clam linguine 199

Cobnuts: wild rice, red cabbage, apple and toasted cobnut salad 113

Cod with lemon, parsley and tomato butter 195

Coffee
coffee jelly with brown bread ice cream 260–61
milk chocolate, almond and espresso fudge 329

Cookies, double chocolate chip 331

Cottage pie, venison with beetroot and apple salad 242–3

Courgettes: griddled courgettes and pine nuts in yoghurt and mint dressing 87

Crab: hot dorset crab on toast 24

Cranberries
venison and cranberry pies 176–7
white chocolate and cranberry fudge 331

Cucumber: chilled tomato, cucumber and fennel soup 50

Curry, chicken and apricot 223

Daylesford Blue, see under blue cheese

Drinks
Daylesford summer wine cup 358–9
lemon refresher 358
seasonal liqueurs 359

Earl Grey cake 315

Eggs; notes on 34
bubble and squeak with fried egg 35–7
corned beef hash cakes 41
garden vegetables with hot Cheddar sauce and salsa verde mayonnaise 15
pan haggerty with mustard, egg and caper mayonnaise 39–40
Rita's baked eggs and onions 43

Elderflower: gooseberry and elderflower jam 365

Energy bars 319

Fennel: chilled tomato, cucumber and fennel soup 50

Feta
- aubergine, pomegranate, feta and pumpkin seed salad 94
- broad bean, pea, mint and feta toasts 16
- tomatoes and feta salad with mint and lemon dressing 81

Fish; notes on 184
- fishcakes, salmon and smoked haddock 190–91
- traditional fish pie 192–3
- see also under individual names

Flowerpot breads 301–5

Fool, gooseberry with shortbread 251

French dressing 335

Fudge
- milk chocolate, almond and espresso 329
- white chocolate and cranberry 331

Game pie 179–81

Garlic
- garlic and chilli dressing 93
- green beans with almonds, parsley and garlic butter 124
- Jerusalem artichoke risotto with garlic and almond breadcrumbs 139–40
- leek and wild garlic pesto risotto 141
- potato wedges with garlic and rosemary butter 126
- purple sprouting broccoli, spelt, crispy garlic and toasted almonds 105
- wild garlic and pumpkin seed pesto 339

Ginger
- ginger biscuits 327
- raw slaw with chilli, soy and ginger dressing 85–6
- rhubarb and ginger jam 364
- salted ginger treacle tart 269

Goat's cheese
- goat's cheese and asparagus tart 156
- griddled butternut squash, goat's cheese and olive salad 93–4

Gooseberries
- gooseberry and elderflower jam 365
- gooseberry fool with shortbread 251

Halibut with Morecambe Bay shrimp butter sauce 196

Ham hock terrine with piccalilli 21

Hazelnuts: pickled pear and hazelnuts with chickpeas, quinoa and Daylesford Blue 97–8

Honey
- butternut squash, honey and sage soup 56
- Manuka honey cake 312
- mushroom, celeriac, truffle honey and toasted pine nuts 110–11
- squash, honey and sage bread 284–5

Hot cross loaf or buns 289–91

Ice cream, brown bread 260–61

Jams 363
- gooseberry and elderflower 365
- red tomato chilli 351
- rhubarb and ginger 364
- strawberry and vanilla conserve 366

Jellies
- coffee jelly with brown bread ice cream 260–61
- orange-poached rhubarb jelly 252
- poached apple and pear jelly with crumble topping and prune cream 254–6

Jerusalem artichokes
- Jerusalem artichoke and cavolo nero tart 164
- Jerusalem artichoke risotto with garlic and almond breadcrumbs 139–40

Kale
- Adlestrop and kale tart 155
- butternut squash and kale tart 163
- chestnut, quinoa, kale and broccoli salad 102–3

Kedgeree, baked salmon, spinach and smoked haddock 200–201

Lamb
- lamb and butter bean casserole with tomatoes, caperberries and olives 213–14
- pressed lamb 219
- slow-cooked lamb shoulder with white beans and salsa verde mayonnaise 215–16

Lasagne, venison and cavolo nero 239

Leeks
- baked leeks with cider 132–3
- chicken, leek and bacon pie 172–3
- leek and potato soup 57
- leek and wild garlic pesto risotto 141

Lemons
- cod with lemon, parsley and tomato butter 195
- lemon drizzle cake 317
- lemon refresher 358
- lemon shortbread biscuits 325
- tomatoes and feta salad with mint and lemon dressing 81
- wilted spinach with toasted pine nuts, sultanas and lemon zest 123

Lentils
- braised brisket with lentils 234
- grilled mackerel with roasted beetroot and spiced lentils 185–6
- lentils, tomato, Daylesford Blue and red onion salad 99

Liqueurs 359

Mackerel
- grilled mackerel with roasted beetroot and spiced lentils 185–6
- smoked mackerel pâté with Daylesford toasts 13

Marmalade, Seville orange 371

Mayonnaise 335
- herb 339
- minted aioli 340
- mustard, egg and caper 39–40
- salsa verde 336
- tuna and caper 83
- watercress 187–9

Meat; notes on cooking 210–11

Mince pies 354–5

Minestrone soup, ten vegetable and two grain 59

Mint
- asparagus, spelt, peas and mint salad 76
- broad bean, pea, mint and feta toasts 16
- chilled pea and mint soup 49
- griddled courgettes and pine nuts in yoghurt and mint dressing 87
- minted aioli 340
- smashed broad beans, peas and mint 121
- tomatoes and feta salad with mint and lemon dressing 81

Mozzarella: grilled peaches, spelt, peas, rocket and mozzarella salad 75

Mulled wine and orange trifle 257–9

Mushrooms
    featherblade of beef with creamed wild mushrooms 230–31
    mushroom, celeriac, truffle honey and toasted pine nuts 110–11
    woodland mushroom shepherd's pie 128–9

Mustard leaves: raw beetroot, kidney beans and mustard leaf salad with horseradish dressing 106

Nettle flowerpot bread 304–5

Olives
    crushed new potatoes with olives, capers and herbs 119
    griddled butternut squash, goat's cheese and olive salad 93–4
    lamb and butter bean casserole with tomatoes, caperberries and olives 213–14

Onions
    balsamic onions 93
    Cheddar, potato and onion pie 169–71
    lentils, tomato, Daylesford Blue and red onion salad 99
    onion purée 126
    red onion, Cheddar and chilli flowerpot bread 302–3
    red onion tarte tatins with Baywell cheese 160–61
    Rita's baked eggs and onions 43
    tomato and sourdough salad with red pepper, onion and basil 82

Oranges
    blood orange and polenta cake with orange whipped cream 272–5
    mulled wine and orange trifle 257–9
    orange-poached rhubarb jelly 252
    seville orange marmalade 371

Pan haggerty with mustard, egg and caper mayonnaise 39–40

Parmesan, chilli and Marcona almond biscuits 324

Pasta
    clam linguine 199
    venison and cavolo nero lasagne 239

Pastry; notes on 152
    savoury 153–4
    sweet 266–7

Pâté, smoked mackerel with Daylesford toasts 13

Peaches; grilled peaches, spelt, peas, rocket and mozzarella salad 75

Pearl barley
    beef, ale and barley casserole 238
    pearl barley, asparagus and pea shoot risotto 145
    ten vegetable and two grain minestrone soup 59

Pears
    pickled pear and hazelnuts with chickpeas, quinoa and Daylesford Blue 97–8
    poached apple and pear jelly with crumble topping and prune cream 254–6

Peas
    asparagus, spelt, peas and mint salad 76
    broad bean, pea and watercress risotto 137
    broad bean, pea, mint and feta toasts 16
    chilled pea and mint soup 49
    garden vegetables with hot Cheddar sauce and salsa verde mayonnaise 15
    grilled peaches, spelt, peas, rocket and mozzarella salad 75
    pearl barley, asparagus and pea shoot risotto 145
    smashed broad beans, peas and mint 121

Peppers
    curried cauliflower, red pepper and nigella seeds 101
    tomato and sourdough salad with red pepper, onion and basil 82

Pesto, watercress or wild garlic and pumpkin seed 339

Pies
    Aunt Mimi's mince pies 354–5
    Cheddar, potato and onion 169–71
    chicken, leek and bacon 172–3
    venison and cranberry 176–7
    woodland mushroom shepherd's pie 128–9
    Wootton Estate game pie 179–81

Piccalilli 343

Pickled vegetables 342–3

Pine nuts
    griddled courgettes and pine nuts in yoghurt and mint dressing 87
    mushroom, celeriac, truffle honey and toasted pine nuts 110–11
    wilted spinach with toasted pine nuts, sultanas and lemon zest 123

Polenta: blood orange and polenta cake with orange whipped cream 272–5

Pollock
    pan-roasted pollock with crushed potatoes and watercress mayonnaise 187–9
    traditional fish pie 192–3

Pomegranates: aubergine, pomegranate, feta and pumpkin seed salad 94

Potatoes
    beetroot, swede and potato bake 131
    bubble and squeak with fried egg 35–7
    Cheddar, potato and onion pie 169–71
    corned beef hash cakes 41
    crushed new potatoes with olives, capers and herbs 119
    leek and potato soup 57
    pan haggerty with mustard, egg and caper mayonnaise 39–40
    pan-roasted pollock with crushed potatoes and watercress mayonnaise 187–9
    potato wedges with garlic and rosemary butter 126
    traditional fish pie 192–3
    venison cottage pie with beetroot and apple salad 242–3
    woodland mushroom shepherd's pie 128–9

Prunes: poached apple and pear jelly with crumble topping and prune cream 254–6

Puddings; notes on 246

Pumpernickel 286–8

Pumpkin: spiced pumpkin, butter bean and spinach casserole 134

Queen of puddings, rhubarb 247–9

Quinoa
    chestnut, quinoa, kale and broccoli salad 102–3
    pickled pear and hazelnuts with chickpeas, quinoa and Daylesford Blue 97–8

Relish, tomato, basil and caper 27

Rhubarb
    orange-poached rhubarb jelly 252
    rhubarb and ginger jam 364
    rhubarb queen of puddings 247–9

Rice
    baked salmon, spinach and smoked haddock kedgeree 200–201
    vanilla rice pudding with apple and blackberry compote 264

wild rice, red cabbage, apple and toasted cobnut salad **113**
see also risotti

Risotti; notes on **136**
broad bean, pea and watercress **137**
Jerusalem artichoke risotto with garlic and almond breadcrumbs **139–40**
leek and wild garlic pesto **141**
pearl barley, asparagus and pea shoot **145**
spelt, garden vegetable and herb **142**

Roasts; notes on **224–5**

Sage
butternut squash, honey and sage soup **56**
squash, honey and sage bread **284–5**

Salads; notes on **70–71**
asparagus, spelt, peas and mint **76**
aubergine, pomegranate, feta and pumpkin seed **94**
beetroot and apple **242–3**
broad bean, bulgar wheat and herb **73**
chestnut, quinoa, kale and broccoli **102–3**
chicken Caesar **114–15**
cold rose veal with tuna and caper mayonnaise **83**
crunchy 'chopped' vegetables **89**
curried cauliflower, red pepper and nigella seeds **101**
griddled butternut squash, goat's cheese and olive **93–4**
griddled courgettes and pine nuts in yoghurt and mint dressing **87**
grilled peaches, spelt, peas, rocket and mozzarella **75**
lentils, tomato, Daylesford Blue and red onion **99**
mushroom, celeriac, truffle honey and toasted pine nuts **110–11**
pickled pear and hazelnuts with chickpeas, quinoa and Daylesford Blue **97–8**
purple sprouting broccoli, spelt, crispy garlic and toasted almonds **105**
raw beetroot, kidney beans and mustard leaf with horseradish dressing **106**
raw slaw with chilli, soy and ginger dressing **85–6**
tomato and sourdough with red pepper, onion and basil **82**
tomatoes and feta with mint and lemon dressing **81**
wild rice, red cabbage, apple and toasted cobnuts **113**

Salmon
baked salmon, spinach and smoked haddock kedgeree **200–201**
salmon and smoked haddock fishcakes **190–91**
traditional fish pie **192–3**

Sardines on toast with tomato, basil and caper relish **27**

Sauces
brown **361**
hot cheddar **341**
tomato ketchup **361**

Shepherd's pie, woodland mushroom **128–9**

Shortbread biscuits, lemon **325**

Shrimps: halibut with Morecambe Bay shrimp butter sauce **196**

Single Gloucester, spinach and smoked bacon tart **168**

smoked haddock
baked salmon, spinach and smoked haddock kedgeree **200–201**
salmon and smoked haddock fishcakes **190–91**
traditional fish pie **192–3**

Soups
beetroot, bacon and crème fraîche **53**
butternut squash, honey and sage **56**
celeriac and apple **47**
chicken (or turkey), ginger and vegetable broth **55**
chilled pea and mint **49**
chilled tomato, cucumber and fennel **50**
leek and potato **57**
ten vegetable and two grain minestrone **59**

Sourdough **296–7**
seven seeds sourdough **299**
tomato and sourdough salad with red pepper, onion and basil **82**
Welsh rarebit and chutney **18–19**

Spelt
asparagus, spelt, peas and mint salad **76**
grilled peaches, spelt, peas, rocket and mozzarella salad **75**
purple sprouting broccoli, spelt, crispy garlic and toasted almonds **105**
spelt, garden vegetable and herb risotto **142**
ten vegetable and two grain minestrone soup **59**

Spinach
baked salmon, spinach and smoked haddock kedgeree **200–201**
Single Gloucester, spinach and smoked bacon tart **168**
spiced pumpkin, butter bean and spinach casserole **134**
wilted spinach with toasted pine nuts, sultanas and lemon zest **123**

Stocks; notes on **372**
chicken **373–4**
vegetable **372**

Strawberry and vanilla conserve **366**

Swede
beetroot, swede and potato bake **131**
crushed, buttered root vegetables and cabbage **127**

Tarts
Adlestrop and kale **155**
apricot and almond **271**
blackberry and apple crumble **268**
Bledington Blue cheese and broccoli **159**
butternut squash and kale **163**
goat's cheese and asparagus **156**
Jerusalem artichoke and cavolo nero **164**
red onion tarte tatins with baywell cheese **160–61**
salted ginger treacle **269**
Single Gloucester, spinach and smoked bacon **168**
three tomato **164–7**

Terrines
ham hock with piccalilli **21**
venison with tomato and chilli jam **22–3**

Toasts; notes on **12**
broad bean, pea, mint and feta toasts **16**
garden vegetables with hot Cheddar sauce and salsa verde mayonnaise **15**
ham hock terrine with piccalilli, **21**
hot dorset crab on toast **24**
sardines on toast with tomato, basil and caper relish **27**
smoked mackerel pâté with Daylesford toasts **13**
venison terrine with tomato and chilli jam **22–3**
Welsh rarebit and chutney **18–19**

Tomatoes; notes on **79**
chilled tomato, cucumber and fennel soup **50**
cod with lemon, parsley and tomato butter **195**

lamb and butter bean casserole with tomatoes, caperberries and olives 213–14
lentils, tomato, Daylesford Blue and red onion salad 99
red tomato chilli jam 351
sardines on toast with tomato, basil and caper relish 27
three tomato tart 164–7
tomato and sourdough salad with red pepper, onion and basil 82
tomatoes and feta salad with mint and lemon dressing 81

Treacle tart, salted ginger 269

Trifle, mulled wine and orange 257–9

Tuna: cold rose veal salad with tuna and caper mayonnaise 83

Turkey, ginger and vegetable broth 55

Veal: cold rose veal salad with tuna and caper mayonnaise 83

Vegetables; notes on 118
chicken (or turkey), ginger and vegetable broth 55
crunchy 'chopped' vegetables 89
crushed, buttered root vegetables and cabbage 127
garden vegetables with hot Cheddar sauce and salsa verde mayonnaise 15
pickled vegetables 342–3
spelt, garden vegetable and herb risotto 142
ten vegetable and two grain minestrone soup 59
vegetable stock 372
see also individual names

Venison
venison and cavolo nero lasagne 239
venison and cranberry pies 176–7
venison cottage pie with beetroot and apple salad 242–3
venison terrine with tomato and chilli jam 22–3

Walnuts: blue cheese and walnut biscuits 323

Watercress
broad bean, pea and watercress risotto 137
pan-roasted pollock with crushed potatoes and watercress mayonnaise 187–9
watercress and pumpkin seed pesto 339

Welsh rarebit and chutney 18–19

Acknowledgements

This book is dedicated to my grandchildren: Tilly, Caspian, Teddy, Scarlet, Atticus, Otis and Iris.

It is a celebration of food by our passionate team here at Daylesford, and a compilation of favourite café and family recipes, and there are so many people to thank for their contributions.

Photographers Sarah Maingot and Martin Morrell for their beautiful photography – both are a joy to work with.

John Hardwick, our trusted chef for many years, who created the recipes. As John says, the food at Daylesford has evolved over the years and many chefs who have worked in the kitchens have contributed to the dishes and recipes. In particular we would like to thank the following people for their inspiration and input: head chef Gaven Fuller and chefs Andy Wheeler; Paul Collins, Kuttiya Nimcean, Chris Webb, Ivan Reid, Alex Holder and Karl MacEwan. A big thank you too to Robin Gosse, Adam Caisley, Annabelle Briggs and Marianne Lumb for fine-tuning all the recipes so carefully for the home kitchen.

Special thanks go to Richard Smith for managing the farm; Jez Taylor and his team for the wonderful produce they grow for us in the market garden; Peter Kindel and his fellow cheese-makers in the creamery, Tibor Kethelyi and Stephen Tarling and their dedicated team of bread and patisserie makers; Tim Field, our head of sustainability, who guides us in our efforts to be as self-sustaining and ethical as is possible; and Nick Fletcher, who steers the ship so brilliantly – with calm and patience and an enormous sense of humour and fun.

Amy Devine, thank you, as always, for your creative vision and sensitivity in reviving these pages and designing the beautiful new cover. Sophie Richardson, I am forever grateful for your unwavering support; thank you for overseeing and managing this project. And Imogen Fortes, my thanks for all that you do in helping me to put my words on the page.

Brooke Litchfield is always a source of inspiration for me. And I would be lost without Michele and Emma, who take care of everything I do – I cannot thank you enough.

My biggest thanks go to the rest of the team at the Mothership: you are the beating heart of Daylesford and I am so fortunate to have your dedication and your passion.

Thanks are also due to all the suppliers and food producers – many small and artisan – who help to make Daylesford what it is today, and to our many friends for their support and enthusiasm, particularly Skye Gyngell, Patrick Holden, Carlo Petrini, Rose Prince and Ruth Rogers.

To our publisher, Rowan Yapp, at Square Peg: thank you for your support and belief in sharing our message and our recipes; and to Shabana Cho and Muna Reyal for the part they've each played in helping to produce the book.

To my children, Alice, Jo and George, for all the help they give me in their own ways.

And finally, a huge thank you to Anthony for his never-ending patience. Without his support none of this would have been possible.

About the author

Carole Bamford has been a champion of sustainable, mindful living for over 40 years, driven by her profound belief that we need to work in harmony with nature, to nurture and protect it and treat it with respect.

As the founder of Daylesford, she is recognised as a visionary in organic farming and food retail. A simple desire to make a small difference to the health and food of her family led her to turn their farmland over to organic, and what began as a collection of empty barns and fields has grown to become one of the UK's most sustainable organic farms.

Bamford was born out of Carole's belief that what we put on our body is as important as what we put into it through our food. Through the clothing, bath, body and homeware collections, she has been an energetic promoter of natural, organic beauty products and sustainably produced garments made from natural fibres.

Carole believes that artisan skills, traditions and craftsmanship should be championed and supported, a belief that in 2019 led to the realisation of a long-standing vision – the opening of Nila House in Jaipur. A centre of excellence, Nila is dedicated to honouring and preserving the natural dye and handloom traditions of India. Its aim is to support artisans across India, to celebrate the art of indigo dyeing and to offer a dynamic platform for cultural and artistic exchange. Nila is part of the Lady Bamford Foundation and is a not-for-profit organisation

In 2019 Carole launched *Seed*, a biannual magazine dedicated to sustainable living, which hopes to gently inspire its readers to be conscious with their choices in order to have a positive impact on the future of our planet.

Daylesford and Bamford have been recognised with numerous awards.

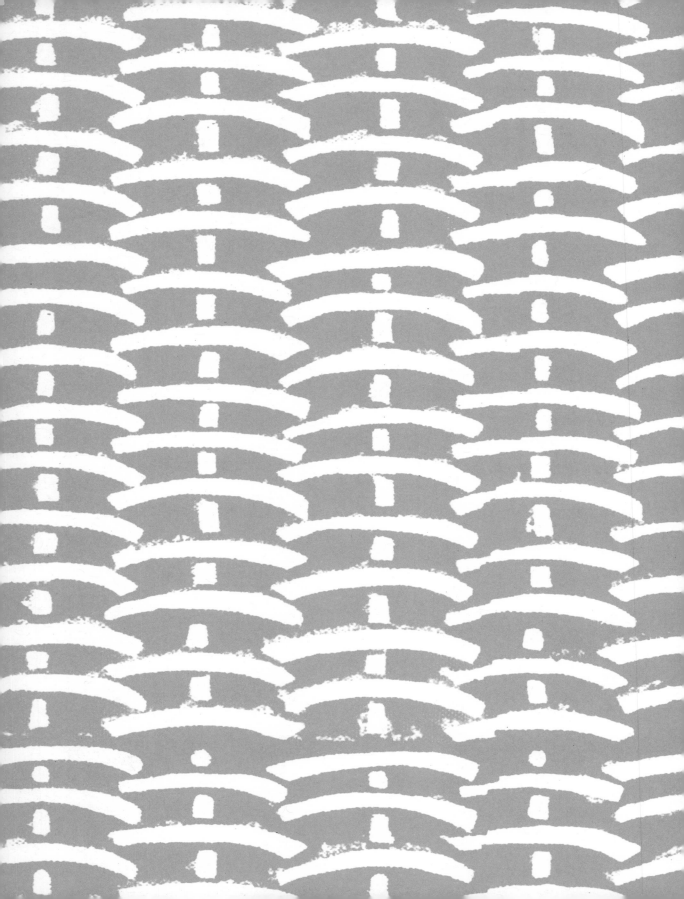